Springerbriefs in Agriculture

For further volumes:
http://www.springer.com/series/10183

P. Srinivasa Rao · C. Ganesh Kumar
Editors

Characterization of Improved Sweet Sorghum Cultivars

Springer

Editors
P. Srinivasa Rao
Sorghum Breeding—ICRISAT
Patancheru, Andhra Pradesh
India

C. Ganesh Kumar
Chemical Biology Laboratory
CSIR-Indian Institute of Chemical
Technology
Hyderabad, Andhra Pradesh
India

ISSN 2211-808X   ISSN 2211-8098   (electronic)
ISBN 978-81-322-0782-5   ISBN 978-81-322-0783-2   (eBook)
DOI 10.1007/978-81-322-0783-2
Springer New Delhi Heidelberg New York Dordrecht London

Library of Congress Control Number: 2012946623

Printed on acid-free paper

Springer is part of Springer Science+Business Media (www.springer.com)

Preface

Food and energy security are the most critical subjects for the sustenance of modern civilization. In view of the depleting oil resources and negative environmental impacts associated with the use of fossil fuels, there is a renewed interest in renewable biofuels, which can form the swivel for sustainable development in terms of socio-economic and environmental considerations. As it is a locally available resource, and there is a possible utilization at the local level, the developing nations are set to gain. Biofuel production and consumption in many developing countries like India, Philippines, Indonesia, Argentina, Nigeria, and Mozambique is at the nascent stage and are evolving. For example, India approved the National Policy on Biofuels on December 24, 2009, which envisaged the use of renewable energy resources as an alternate fuel to supplement transport fuels and proposed an indicative target of 20 % blending of biofuels by 2017. The bioethanol blending programme followed in India has two major bottlenecks; first, there is lack of sufficient ethanol for blending, and second, the purchase price of ethanol is low. It is anticipated that sooner these two issues will be addressed. The biofuel policy also identified sweet sorghum as a unique biofuel feedstock due to its potential to supply food, feed, and fodder simultaneously.

The International Crops Research Institute for the Semi-Arid Tropics (ICRISAT) has a BioPower initiative that aims to empower the dryland poor to benefit from, rather than be marginalized by, the biofuels revolution. The Institute has developed diverse and improved sweet sorghum varieties, female hybrid parents, and hybrids that can be used as feedstock sources for bioethanol production. The fodder quality is good and nutritious for livestock. The Indian National Agricultural Research System (NARS) started working on sweet sorghum to produce sugar in mid-1980s to augment sugar production. However, this effort could not succeed due to crystallization problems with sweet sorghum juice. Since then many varieties like SSV 74, SSV 84, and RSSV 9 and also a hybrid, CSH 22SS were released to explore for bioethanol production as well as for fodder use.

This book on the characterization of tropical sweet sorghum chronicles sweet sorghum history, comparative performance with other competing feedstocks, breeding efforts, morpho-biochemical traits of rainy and post-rainy season adapted cultivars (supported by colored photographs for easy identification) besides the status of commercialization which is described in five distinct chapters.

Written by highly experienced scientists from ICRISAT and CSIR-Indian Institute of Chemical Technology (CSIR-IICT), this lucid and comprehensive publication is a valuable source of information on the genesis and progress of sweet sorghum research and morpho-biochemical traits of tropical sweet sorghums. It will serve as an important source of reference to researchers, students, entrepreneurs, policymakers, and other stakeholders in India and in many developing countries as well.

P. Srinivasa Rao

C. Ganesh Kumar

Contents

About the Book

A number of driving forces, including the soaring global crude oil prices and environmental concerns in both developed and developing nations has triggered a renewed interest in the recent years on the R&D of biofuel crops. In this regard, many countries across the globe are investing heavily in the bioenergy sector for R&D to increase their energy security and reduce their dependence on imported fossil fuels. Currently, most of the biofuel requirement is met by sugarcane in Brazil and corn in the United States, while biodiesel from rapeseed oil in Europe. Sweet sorghum has been identified as a unique biofuel feedstock in India since it is well adapted to Indian agro-climatic conditions and more importantly it does not jeopardize food security at the cost of fuel.

Sweet sorghum [*Sorghum bicolor* (L.) Moench] is considered as a SMART new generation energy crop as it can accumulate sugars in its stalks similar to sugarcane, but without food-fuel trade-offs and can be cultivated in almost all temperate and tropical climatic conditions and has many other advantages. The grain can be harvested from the panicles at maturity. There is no single publication detailing the agronomic and biochemical traits of tropical sweet sorghum cultivars and hybrid parents. Hence, an attempt is made in this publication—"Characterization of improved sweet sorghum cultivars" to detail the complete description of cultivars. This book serves as a ready reference on the detailed characterization of different improved sweet sorghum genotypes following the PPVFRA guidelines for the researchers, entrepreneurs, farmers, and other stakeholders to identify the available sweet sorghum cultivars and understand their yield potential in tropics.

Sweet Sorghum: From Theory to Practice

P. Srinivasa Rao, C. Ganesh Kumar and Belum V. S. Reddy

Abstract Sweet sorghum [*Sorghum bicolor* (L.) Moench] is a multipurpose crop (food, feed, fodder and fuel) that has the potential as an alternative biofuel feedstock without impacting food and fodder security. This chapter entitled "Sweet sorghum: From theory to practice" discusses on the historical developments in sweet sorghum and immense range of genetic variability that was available in major sorghum regions of the world. The candidate feedstock characteristic traits of sweet sorghum *vis-a-vis* other major biofuel feedstocks like sugarcane, corn and sugar beet were compared. Sweet sorghum fares well in many aspects as it is a C_4 species with greater resilience to diverse agro-ecologies, low fertilizer and water requirement besides short lifecycle. Hence, many consider it as climate change ready crop; some consider it as miracle crop and few term it as a smart crop. A quantitative insight into the production-ecological sustainability of sweet sorghum biofuel feedstock production systems has been discussed. The ongoing R&D efforts at ICRISAT as well as in National Agricultural Research System (NARS) on sweet sorghum value chain were highlighted. The breeding efforts in Brazil, USA and China on this crop are briefly narrated.

Keywords Sweet sorghum · Biofuels · Semi-arid tropics · C_4 plant · Jowar · Agronomy · Taxonomy, food—fuel trade off

P. Srinivasa Rao (✉) · B. V. S. Reddy
International Crops Research Institute for the Semi-Arid Tropics (ICRISAT),
Patancheru 502324, India
e-mail: psrao72@gmail.com

C. Ganesh Kumar
Chemical Biology Laboratory, CSIR-Indian Institute of Chemical Technology (CSIR-IICT),
Uppal Road, Hyderabad 500607, India

P. S. Rao and C. G. Kumar (eds.), *Characterization of Improved Sweet Sorghum Cultivars*, SpringerBriefs in Agriculture,
DOI: 10.1007/978-81-322-0783-2_1, © The Author(s) 2013

1 Introduction

Sorghum (*Sorghum bicolor* L. Moench) is a C_4 herbaceous annual grass that is cultivated from the seed, and is known by various names like great millet and guinea corn in West Africa, kafir corn in South Africa, durra in Sudan, mtama in Eastern Africa, jowar in India, kaoliang in China and milo or milo-maize in the United States (Purseglove 1972). It has wide flat leaves and a round or elliptical panicle with full of grain at maturity. The plant accumulates high concentrations of soluble sugars (10–15 %) in the plant stalk sap or juice. It is a crop of high universal value since it can be cultivated in tropical, subtropical, temperate, and semi-arid regions as well as in poor quality soils of the world. It is termed as "the sugarcane of the desert" or "the camel among crops" due to its drought hardy characteristics (Sanderson et al. 1992).

The name "sweet sorghum" is used to identify those varieties of sorghum, *Sorghum bicolor* (L.) Moench, which has juicy and sweet stalks. Sweet sorghum is mainly cultivated for syrup production or forage, whereas other sorghum varieties such as kafirs and milos are cultivated for grain. Other sorghum-types include the broomcorn-type sorghum (*Sorghum dochna* var. *technicum* (Koern.) Snowden), whose head/panicle is used for making brooms and brushes; while johnsongrass, *Sorghum halepense* (L.) Pers. and sudangrass, *Sorghum sudanense* (Piper) Stapf. are grown primarily for forage purpose.

2 History

Sorghum is a grass of Old World origin and *Sorghum vulgare* Pers. is a native wild plant of Africa that is drought-resistant and heat-tolerant member of the grass family and many of the varieties under cultivation in the recent history have originated from that continent. Documented evidence indicated that sorghum was grown in Assyria as early as 700 BC (Ziggers 2006). Wide varieties in the genus *Sorghum* were observed in the North Eastern regions of Africa comprising of Ethiopia and Sudan in Eastern Africa (Doggett 1988). Around 200 AD or even earlier, sorghum made its way into Eastern Africa from Ethiopia via the local tribes, who cultivated this crop mainly for grain and the sweet cane was chewed for pleasure and nutrition. Later, the Bantu tribe carried this crop with them to the Savannah regions of Western and Southern Africa who used the grain mainly for making beer. The Bantu tribe later moved this crop during their expansion from Southern Cameroon region around first century AD, and the southern border of the Congo forest belt. The present-day sorghums of Central and Southern Africa bear close relationship with those of the Tanzania and are more distantly related to those of West African varieties, since the equatorial forests were an effective barrier to their spread (FAO 1995).

During the first millennium BC, sorghum was probably carried to India from Eastern Africa in ships as food by the chow traffic which operated for about

3000 years between East Africa (the Azanean Coast) and India via the Sebaean Lane in Southern Arabia. The sorghum varieties of India bear relationship to those existing in Northeastern Africa and the coast between Cape Guardafui and Mozambique. This crop might have spread along the coast of Southeast Asia and around China around the beginning of Christian era; however, a possibility that cannot be denied is that sorghum might have arrived much earlier in China by the silk trade routes (FAO 1995). Later it made its way to Western parts of the World via Asia. This plant was mentioned in European botanical literature in 1542 and was referred to as Sorghi, the name similar to that used in India.

Sorghum was introduced in the Caribbean Islands and other Latin American countries from West Africa through the slave trade and by navigators plying the Europe-Africa-Latin America trade route in the early 17th century as another source for sugar production. The case is similar for Australia. These early varieties were established as "guinea corn" (also called as Rural Branching Durra). However, the guinea corn in course of time disappeared from production. The tropical adapted varieties were introduced via slave trader ships as broomcorn variety by Benjamin Franklin in the United States in 1725, while Johnson grass was introduced as forage grass variety in South Carolina in 1830. These varieties were cultivated extensively in the US after 1850s, when sweet sorghum was introduced in 1853 by William Prince, a New York nurseryman who received some seed from France via China and cultivated the sorghum crop in New York. He claimed that sweet sorghum was a potential new sugar crop and sold the seed to farmers around Northern America for mass cultivation. In a parallel effort, J.D. Browne, a United States patent agent, traveled to France and noticed French efforts on sweet sorghum cultivation for sugar production from the sweet canes, which grew in places having similar climatic conditions favoring corn cultivation. Based on these observations, Browne collected seed from France and sent them back to the US Patent office and advised that this crop can act as a new sugar source and could be cultivated in America's Corn Belt like American North and Midwest regions.

The sweet sorghum varieties introduced by William Prince and J.D. Browne were termed as "Black amber" or "Chinese sugarcane" since they arrived in America though France via China. The Chinese sugar cane variety was also known as Eusorghum. Since then many sweet and forage varieties were introduced in the US from China, Africa and Australia and were domesticated (Vinall et al. 1936; Ziggers 2006). Subsequently, sorghum production was established in the United States to a larger extent with the introduction of grain sorghum variety in California in 1874 and the milo variety's introduction by the Colombian missionary H.B. Pratt in 1879. In the early 1900s, grain sorghum was identified as a drought-tolerant crop and its production surpassed corn in the arid regions of the Southern Great Plains. Scientists from various agricultural experiment stations and USDA scientists from Texas, Oklahoma and Kansas recorded the sorghum's drought tolerance and with the help of seed production farmers selected improved phenotypes. Many local land laces of sweet stalked sorghum found in Western and Central Africa (Mali, Niger and Tanzania) are used for staple purpose.

3 Sweet Sorghum and its Utilization

Characteristics: The term sweet sorghum is used to distinguish varieties of sorghum with high concentrations of soluble sugars in the plant stalk sap or juice compared to grain sorghum which has relatively less sugar and juice in the stalks. Sweet sorghum is a C_4 plant species having wide flat leaves and a round or elliptical head with full of grain at the stage of maturity. It is, like grain sorghum, traditionally under cultivation for nearly 3000 years. It can be grown successfully in semi-arid tropics, where other crops fail to thrive and are highly suitable for cultivation in harsh dryland growing areas. With irrigation, it can produce very high yields. During very dry periods, sweet sorghum can go into dormancy, with growth resuming when sufficient moisture levels return (Gnansounou et al. 2005). It can be grown easily on all continents, in tropical, sub-tropical, temperate, semi-arid regions as well as in poor quality soils. It is known as the sugarcane of the desert and also "the camel among crops" for its drought hardy characteristics (Sanderson et al. 1992). It has higher drought tolerance and water use efficiency (WUE) compared to maize, and yields, like those of *Miscanthus*, ranging from 18 to 36 dry t ha^{-1} of biomass per year on low-quality soils with minimal inputs of fertilizer and water. In Indiana, studies showed that sweet sorghum cultivars produced 25–40 tons of dry mass per hectare with 0–60 kg ha^{-1} of nitrogen fertilizer. The high WUE and low N requirements of sorghum also provide significant advantages to the growers, because sorghum fits into a normal rotation scheme with corn and soybeans, yet has lower production costs and employs similar production equipment. Its ratooning ability enables multiple harvests per season, a feature that could expand the geographical range of sorghum cultivation. The grain, stalk juice and bagasse (the fibrous residue that remains after juice extraction) can be used to produce food, fodder, ethanol and power. Owing to these favorable attributes, William D Dar, refers to it as a **SMART** crop (Fig. 1). It's candidate traits *vis-a-vis* utilizable options are listed in Table 1.

These important characteristics, along with its suitability for seed propagation, mechanized crop production, and comparable ethanol production capacity *vis-a-vis* sugarcane and sugarbeet makes sweet sorghum a viable alternative source for ethanol production (Table 2).

The sweet sorghum value chain basically involves four critical areas i.e. feed stock supply, sugars conversion, bioenergy (ethanol blended gasoline) distribution and use (Fig. 2). In a feedstock like sweet sorghum, whole plant, its products and byproducts are used for diverse purposes.

4 Sorghum Distribution and Climatic Conditions

Sorghum (*Sorghum bicolor* (L) Moench) is the fifth important cereal crop in the world in production and fifth in acreage after wheat, rice, maize and barley. It is mostly grown in the semi-arid tropics (SAT) of the world wherein the production

Fig. 1 An ICRISAT improved sweet sorghum variety, ICSV 25274

system is constrained by poor soils, low and erratic rainfall and low inputs resulting in low productivity. In terms of area, India (7.5 m ha) is the largest sorghum grower in the world followed by Nigeria (7.6 m ha) and Sudan (6.6 m ha). India is the third largest producer after USA and Nigeria. Sorghum is well adapted to the SAT and is one of the most efficient dryland crops to convert atmospheric CO_2 into sugar (Srinivasa Rao et al. 2009). The crop can be grown in a wide range of climatic conditions as given below.

Latitude: Sorghum is grown between 45°N and 45°S latitude on either side of the equator.

Altitude: Sorghum can be found at elevations between mean sea level and 1,500 m. Most East African sorghum is grown between the altitudes of 900–1,500 m, and cold-tolerant varieties are grown between 1,600 and 2,500 m in Mexico.

Temperature: Sweet sorghum can be grown in the temperature range of 12–37 °C and optimum temperature for growth and photosynthesis is 32–34 °C, day length is 10–14 h, optimum rainfall ranges from 550 to 800 mm and relative humidity ranges between 15 to 50 %.

Soils: Alfisols (red) or vertisols (black clay loamy) with pH ranging between 6.5 to 7.5, organic matter >0.6 %, depth >80 cm, bulk density <1.4 gcc, water holding capacity >50 % field capacity, N $\geq$ 260 kg ha^{-1} (available), P $\geq$ 12 kg ha^{-1} (available), K $\geq$ 120 kg ha^{-1} (available).

Water: Sorghum will survive with a supply of less than 300 mm over the season of 100 days, while it responds favorably with additional rainfall or

Table 1 Candidate traits of sweet sorghum as biofuel feedstock (Reddy et al. 2005; Srinivasa rao et al. 2009, 2010)

As crop	As ethanol source	As bagasse	As raw material for industrial products
Short duration (3–4 months)	Amenable to eco-friendly processing	High biological value	Cost-effective source of pulp for paper making
C_4 dryland crop	Less sulphur in ethanol	Rich in micronutrients	Dry ice, acetic acid, fusel oil and methane can be produced from the co-products of fermentation
Good tolerance of biotic and abiotic constraints	High octane rating	Use as feed, for power co-generation or bio-compost	
Meets fodder and food needs	Automobile friendly (up to 25 % of ethanol-petrol mixture without engine modification)	Good for silage making	Butanol, lactic acid, acetic acid and beverages can be manufactured
Non-invasive species			
Low soil N_2O and CO_2 emission			
Seed propagated			

irrigation water. Typically, sweet sorghum needs between 500 to 1000 mm of water (rain and/or irrigation) to achieve good yields, i.e., 50–100 t ha^{-1} total above ground biomass (fresh weight). Though sorghum is a dryland crop, sufficient moisture availability for plant growth is critically important for high yields. The major advantage of sorghum is that it can become dormant especially in vegetative phase under adverse conditions and can resume growth after relatively severe drought. Early drought stops growth before panicle initiation and the plant remains vegetative; it will resume leaf production and flower when conditions again become favorable for growth. Mid-season drought stops leaf development. Sorghum is susceptible to sustained flooding, but will survive temporary water logging much better than maize.

Radiation: Being a C_4-plant, sweet sorghum has high radiation use efficiency (RUE, about 1.3–1.7 g M J^{-1}). It has been shown that taller sorghum types possess higher RUE, because of a better light penetration in the leaf canopy.

Photoperiodism: Most hybrids of sweet sorghum are relatively less photoperiod-sensitive. Traditional farmers, particularly in West Africa, use photoperiod-sensitive varieties. With photoperiod-sensitive types, flowering and grain maturity occur almost during the same calendar days regardless of planting date, so that even with delayed sowing, plants mature before soil moisture is depleted at the end of the season.

5 Taxonomy

The name *Sorghum bicolor* (L.) Moench was proposed by Clayton in 1961 as the correct name for the cultivated sorghum which is currently in use (Spangler 2003). The genus *Sorghum* is a variable complex genus belonging to the tribe

Table 2 Comparison of sweet sorghum with other bioethanol feedstocks (Reddy et al. 2008; Srinivasa Rao et al. 2009; Almodares and Hadi 2009; Wortmann et al. 2010; Girase 2010)

Characteristics	Sugarcane	Sugar beet	Corn	Sweet sorghum
Crop duration (months)	12–13	5–6	3–4	4
Growing season	One season	One season	All seasons	All seasons (if water is available)
Propagation	Seed (40,000 ha^{-1})	Seed (3.6 kg ha^{-1}; pellet)	Seed (25 kg ha^{-1})	Seed (8 kg ha^{-1})
Soil requirement	Grows well in drained soil	Grows well in sandy loam; also tolerates alkalinity		All types of drained soil
Water management	Requires water throughout the year (36,000 m^3 ha^{-1})	Requires water, 40–60 % compared to sugarcane (18,500 m^3 ha^{-1})	Requires water (12,000 m^3 ha^{-1})	Less water requirement; can be grown as rain-fed crop (8,000 m^3 ha^{-1})
Crop management	Requires good management 250 to 400 N 125P-125 K	Requires moderate management 120 N-60P-60 K	Requires good management 130 N-60P-60 K	Easy management; low fertilizer 90 N-40 P
Stalk/beet/grain yield (t ha^{-1})	60–85	85–100	5–10	45–65
Sugar content on weight basis	10–12 %	15–18 %		7–12 %
Sugar yield (t ha^{-1})	5–12	11.25–18		3–7
Ethanol yield from juice (l ha^{-1})	4,350–7,000	7,100–10,500	2,150–4,300	2,475–3,500
Harvesting	Harvested mechanically	Harvested mechanically	Harvested mechanically	Very simple; Predominantly manual and mechanical harvesting at pilot scale

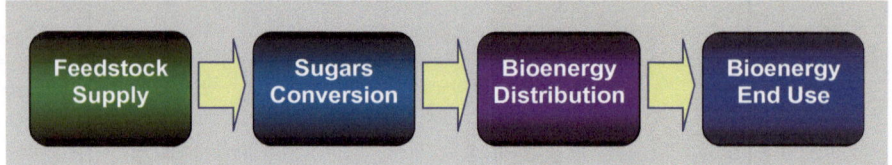

Fig. 2 Critical areas in sweet sorghum ethanol value chain (Srinivasa Rao et al. 2009)

Andropogoneae of the grass family Poaceae, which comprises of 24 species with various chromosome numbers and are subdivided into five sub-generic sections based upon node, panicle and spikelet morphology: Eu-sorghum (Stapf), Chaeto-sorghum, Heterosorghum, Para-sorghum (Snowden) and Stiposorghum (Garber 1950; de Wet 1978; Lazarides et al. 1991; USDA-ARS 2006). Section Eu-sorghum has six species consisting of cultivated, proginetor and weed species such as *Sorghum bicolor, S. arundinaceum, S. drummondii, S. halepense, S. propinquum* and *S. almum*, that have their natural range spread throughout Africa and Asia (de Wet 1978; Duvall and Doebley 1990). The other *Sorghum* species within the section Eu-sorghum forms the secondary gene pool and breeders have accessed genes from these species by introgression. The rest nineteen *Sorghum* species belonging to sub-generic sections other than Eu-sorghum formed the untapped tertiary gene pool. They are distributed primarily in Africa, Asia, Australia and South America, consisting of the cultivated species, their progenitors, and some serious weed species, and have close genetic relationships and inter-crossabilities (de Wet and Harlan 1971; Doggett 1976). *Sorghum bicolor* is a perennial diploid (2n = 20), which further includes three subspecies, namely, *S. bicolor* (cultivated sorghum) and its nearest wild relatives, *S. arundinaceum* (Desv.) de Wet et Harlan (wild sorghums) and *S. drumondii* (Steud.) de Wet (weedy sorghums). Subspecies *bicolor* includes all cultivated races and they are further subdivided into basic and intermediate races. The five basic races include *bicolor, guinea, caudatum, kafir* and *durra* and the ten intermediate races are those between any two of those types, classified primarily based on grain shape, glumes and panicle (Harlan and de Wet 1972).

6 Reproductive Biology

Sorghum is considered as a predominantly self-pollinated species but with cross pollination occurring to an extent of 4–10 % under specific conditions.

Panicle initiation: Sorghum is a short day plant, and blooming is hastened by short days and long nights. However, varieties differ in their photoperiod sensitivity (Quinby and Karper 1947). Tropical sweet sorghum varieties initiate the reproductive stage when day lengths return to 12 h. Floral initiation takes place 30–40 days after germination. Usually, the floral initial is 15–30 cm above the ground when the plants are about 50–75 cm tall (House 1980). Floral initiation marks the end of the vegetative growth due to meristematic activity. The time required for transformation from the vegetative apex to reproductive apex is largely influenced by genetic characteristics and the environment (photo-period and temperature). The grand period of growth in sorghum follows the formation of a floral bud and consists largely of cell enlargement. Hybrids take less time to reach panicle initiation and are relatively less influenced by photo-period and temperature (Srinivasa Rao et al. 2009).

Panicle emergence: During the period of rapid cell elongation, floral initial develops into an inflorescence. About 6–10 days before flowering, the boot will form as a bulge in the sheath of the flag leaf. This will occur, in a variety that flowers in 60–65 days, about 55 days from germination. Sorghum usually flowers in 55 days to more than 70 days in warm climates, but flowering may range from 30 days to more than 100 days. These observations are valid for tropical sweet sorghums, while temperate sorghums that mature in 5 months or later take longer period by 20–30 days for panicle emergence.

Panicle structure: Inflorescence is a raceme, which consists of one or several spikelets. It may be short, compact, loose or open, composed of a central axis that bears whorls of primary branches on every node. The racemes vary in length according to the number of nodes and the length of the internodes. Each primary branch bears secondary branches, which in turn bear spikelets. The central axis of the panicle, the rachis, is completely hidden by the density of the panicle branches in some, while it is completely exposed in others. The spikelet usually occurs in pairs, one being sessile and the second borne on a short pedicel, except the terminal sessile spikelet, which is accompanied by two pediceled spikelets. On the pediceled spikelet, the pedicels vary in length from 0.5 to 3.0 mm, and usually are very similar to the internodes. The first and second glumes of every spikelet enclose two florets; the lower one is sterile and is represented by a lemma, and the upper fertile floret has a lemma and palea. Two lodicules are placed on either side of the ovary at its base. Androecium consists of one whorl of three stamens. The anthers are attached at the base of the ovule by a very fine filament and are versatile and yellowish in color. Gynoecium is centrally placed and consists of two pistils with one ovule from which two feathery stigmas protrude. Many of these floral characters, such as anther color, stigma color, stigma length, length of pedicel, etc. are important traits for cultivar identification and classification.

Sessile spikelets: The sessile spikelet contains a perfect flower. It varies in shape from lanceolate to almost rotund and ovate and is sometimes depressed in the middle. The color is green at flowering, which changes to different colors like straw, cream, yellow, red, brown, purple, or almost black at grain maturity. The intensity and extent of coloring on the glumes is variable. Glumes vary from quite hairy to almost hairless. The seed may be enclosed by the glume or may protrude from it, being just visible to almost completely exposed.

Pediceled spikelets: These are much narrower than the sessile spikelets, usually lanceolate in shape. They can be smaller, the same size, or longer than the sessile spikelets. They possess only anthers but occasionally have a rudimentary ovary and empty glumes.

Anthesis and pollination: Anthesis starts after panicle emergence. Flowers begin to open 2 days after full emergence of the panicle. Floret opening or anthesis is achieved by swelling of the lodicules, and is followed by the exertion of anthers on long filaments and of stigmas from between the lemma and palea. Sorghum head begins to flower at its tip and flowers successively downward over a 4 or 5 day period. Flowering takes place first in the sessile spikelets from top to bottom of the inflorescence. It takes about 6 days for completion of anthesis in the panicle with maximum flowering at 3 or 4 days after anthesis begins. Flowering proceeds downwards to the base in a horizontal plane on the panicle. When flowering of the sessile spikelets is halfway down the panicle, pedicellate spikelets start to open at the top of the panicle and proceed downwards. The flowering phase of pedicellate spikelets overtakes the flowering phase of sessile spikelets before they reach the base of the inflorescence (Maiti 1996). Anthesis takes place during the morning hours, and frequently occurs just before or just after sunrise, but may be delayed on cloudy damp mornings. It normally starts around midnight and proceeds up to 10:00 h depending on the cultivar, location and weather. Maximum flowering is observed between 6:00 and 8:00 h. Because all heads in a field do not flower at the same time, pollen is usually available for a period of 10–15 days. At the time of flowering, the glumes open and all the three anthers fall free, while the two stigmas protrude, each on a stiff style. The anthers dehisce when they are dry and pollen is blown into air. Pollen in the anthers remains alive for several hours after pollen shedding. Flowers remain open for 30–90 min. Dehiscence of the anthers for pollen diffusion takes place through the apical pore. The pollen drifts to the stigma, where it germinates; the pollen tube, with two nuclei, goes down the style, to fertilize the egg and form a 2n nucleus. Glumes close shortly after pollination, though the empty anthers and stigmas still protrude (except in the long-glumed types). The florets of some of the very long-glumed types do not open for fertilization—a phenomenon known as cleistogamy.

Cytoplasmic male sterility has been found in sorghum (A_1–A_4 systems) and has made possible the development of a hybrid seed industry. A good male-sterile plant will not develop anthers, but in some instances dark-colored shriveled anthers with no viable pollen will appear. Partially fertile heads are also observed, and although the anthers frequently have viable pollen, the quantity is less than in normal plants.

Seed: The ovule begins to develop as a light green, almost cream-colored sphere; after about 10 days it begins to take size and becomes darker green. Maturity of grain follows a similar pattern to that of flowering. The development of grains follows a sequence of stages comprising milky, soft dough, hard dough to the final physiological maturity, when a black layer is formed at the hilar region due to the formation of callus tissue. It takes about 30 days for the seeds to reach maximum dry weight (physiological maturity). The seeds contain about 30 % moisture at physiological maturity; they dry to about 10–15 % moisture during the following 20–25 days (House 1980). The seed can be harvested at any time from physiological maturity to seed dryness; however, seed with more than 12 % moisture must be dried before storage. The seeds harvested and dried at physiological maturity have good quality and fetches higher market price (Audilakshmi et al. 2005). There is a distinct varietal difference in the rate of senescence of remaining leaves. All leaves may be dried, or almost dried, at grain maturity; or the plant may remain green. Seed size varies from very small (less than 1 g/100 seeds) to large (5–6 g/100 seeds).

7 Food: Fuel Trade off

It is often stated that sweet sorghum cultivars do not produce grain yield or the grain yield is very less *vis-a-vis* grain sorghum. Studies at ICRISAT showed that sweet sorghum hybrids had higher stem sugar yield (11 %) and higher grain yield (5 %) as compared to grain sorghum types, while sweet sorghum varieties had 54 % higher sugar yield and 9 % lower grain yield as compared with non-sweet stalk varieties in the rainy season. On the other hand, both sweet sorghum hybrids and varieties had higher stalk sugar yields (50 and 89 %) and lower grain yields (25 and 2 %) in the post-rainy season. Thus, there is little tradeoff between grain and stalk sugar yields in the sweet sorghum hybrids in the rainy season, while the trade off is less in varieties in the post-rainy season (Srinivasa rao et al. 2009, 2010; Kumar et al. 2010). The experimental data on the relationship between stalk sugar traits and grain yield shows that the regression coefficient of stalk sugar yield on grain yield is not significant; thereby indicating that the grain yield is not affected when selection is done for stalk sugar yield. Hence selection programs can aim to improve both the traits simultaneously.

8 Crop Agronomy

The already standardized agronomic practices for grain sorghum are not entirely applicable to sweet sorghum because sweet sorghums produce more biomass along with sugars. Developing improved eco-region specific agro-technology and pre- and post-harvesting stalk juice quality studies are the urgent priority. Moreover, the commercial viability of industry hinges upon raw material (sweet sorghum) availability for most part of the year. The adaptation (general and specific) of improved cultivars to different regions and seasons needs to be identified owing to high GGE interaction of sugar yield (Srinivasa Rao et al. 2011a) and its competent traits as described earlier. Standardization of optimized spacing (45 × 15 cm/ 60 × 15 cm/75 × 15 cm), fertilizer application (80–100 kg N, 30–50 P_2O_5), intercultural operations (thinning, weeding, soil mulch), irrigation schedule (both Alfisols and Vertisols apart from seasons), harvest timing and methodology will greatly enhance the productivity of sweet sorghum. In some areas response to micronutrients (like B, Zn and S) in juice yield and quality was observed (Srinivasa Rao et al. 2011b). The crop, even if uptakes different amount of nitrogen, seems to be insensitive to the mineral nitrogen supply and also seems to have a great potentiality in semi-arid environment in terms of yield production (Cosentino et al. 2012). The grain and sugar yields are best in the rainy and summer seasons, whereas in the post-rainy season the grain yield is high, but with less stalk and sugar yield. However, the results from tropical and temperate crosses have resulted in development of few post-rainy season cultivars at ICRISAT. In Brazil, efforts are being made to grow sweet sorghum in a period where stalks are harvested before and after sugarcane season so as to extend the period of operation of distillery. The present day multi-feedstock distilleries can successfully run on a variety of feedstocks like sugarcane, sweet sorghum, cassava and sugarbeet, etc. Agronomic and physiological measures aiding in increasing the period of industrial utilization (PIU) of sweet sorghum (e.g., customized fertilizer application, irrigation at physiological maturity, spraying gibberelic acid (GA), ethrel, solubar, etc., or soil application of micronutrients or other amendments to delay maturity, etc.) will further strengthen the use of sweet sorghum as a biofuel/industrial crop. Rapid sugar accumulation immediately after flowering and its retention for a longer period for staggered feedstock supply is another area of research that deserves immediate attention.

9 R & D Efforts

The USA varieties such as Keller, BJ248, Rio and Wray are some of the popular sweet sorghum varieties grown in the Americas, Europe, China and Thailand. Considerable progress has been made in breeding for improved sweet sorghum lines with higher millable cane and juice yields in India, China, Brazil and several other countries. The Sorghum Institute, Liaoning Academy of Agricultural Sciences has successfully bred and released new sweet sorghum hybrids Liaosiza No.1 in 1989 and Liaosiza No.2 in 1995, which are widely grown throughout China. Similarly in Brazil, Embrapa has released sweet sorghum cultivars like BR501, BR503 BR505, BRS506 BR507 and BRS601 for large scale cultivation (Schaffert personal communication). Dale, Theis, Cowley, Tracey, BJ 248 and Sugardrip are some of the other sweet sorghum varieties grown all over the world.

Sweet sorghum research at ICRISAT was initiated in 1980 to identify lines with high stalk-sugar content in part of the sorghum world germplasm collection maintained at ICRISAT's gene bank initially by chewing the stalks at maturity. Seventy accessions that tasted sweet were evaluated during the rainy season of 1980 and nine accessions with high Brix% were planted again in 1981 rainy season, of which two cultivars, IS-6872 and IS-6896, were selected. The mean Brix% of the nine accessions grown in 1980 and 1981 varied by only about 3 % between the two seasons, indicating that the differences between growing seasons had little influence on the stalk-sugar content. Apart from this, several sweet sorghum lines with high Brix% values were identified among Nigerian lines, Zimbabwe lines, and advanced breeding progenies at ICRISAT-Patancheru. Due to changed focus driven by donor perceptions and National Agricultural Research Systems (NARS) needs, sweet sorghum research at ICRISAT was discontinued in late 1990's. However, ICRISAT renewed its sweet sorghum research to contribute its share to meet the increased demand created for ethanol following the Indian Government's policy to blend petrol and diesel with ethanol and initiate a program for the identification/development of sweet-stalked and high biomass sorghum hybrid parents and varieties during 2002. In an effort to identify promising sweet stalk hybrid parents from the existing diverse set of grain sorghum hybrid parents at ICRISAT, as a short term strategy for immediate utilization for hybrid cultivar development, several B-lines, R-lines and varieties were evaluated for stalk sugar content over the seasons and years. As many as 30 A/B-lines and 35 R-lines/ varieties were found to be better combiners for agronomic and sugar yield related traits. The breeding strategy of ICRISAT revolves around developing hybrid parental lines particularly in partnership with NARS partners. Apart from the above, 27 B-lines and 68 R-lines for rainy season adaptation and 19 B-lines and 35 R-lines for post-rainy season adaptation are in the pipeline. The hybrid cultivar, CSH 22SS, developed and released in India by Indian National Program, had the female parent from ICRISAT. The ICRISAT variety, ICSV 93046, was tested in All India Coordinated Sorghum Improvement Project (AICSIP), Hyderabad from 2005 to 2007 and was found superior to the control varieties (SSV 84 and CSV 19SS) and recommended for identification. Other promising lines from ICRISAT in AICSIP over last 2 years are: ICSV 25274 and ICSSH 58 in India. In 2011, CSV 24SS another sweet sorghum variety bred by Directorate for Sorghum Research (DSR), Hyderabad was released for cultivation. Thousands of hybrids and segregating populations are under evaluation for stalk sugar traits. Research experience at ICRISAT and elsewhere has showed that hybrids produce relatively higher biomass, besides being earlier and more photo-insensitive when compared to the varieties grown under normal as well as abiotic stresses including water-limited environments. The requirement of photo- and thermo-insensitiveness is essential to facilitate plantings at different dates for continuous supply of stalks to distilleries for ethanol production. Therefore, the development of sweet sorghum hybrids is receiving high priority to produce more feedstock and grain yield per drop of water and unit of energy invested.

Increased demand for food triggered by the fast-growing human population, the need to sustain biodiversity, and the spurt in investments in agricultural research by private sector have resulted in seeking the Intellectual Property Rights (IPR) for the valuable research products. This has led to the introduction of plant variety protection legislations across the world particularly in European countries and in the USA. International efforts to harmonize the IPRs across countries to improve trade led to holding of various conventions [including Union Pour la Protection des Obtentions Végétales (UPOV) 1991], leading to the establishment of guidelines on Plant Breeder's Rights (PBRs). This was followed by Uruguay round of deliberations, resulting in Trade Related Intellectual Property (TRIPs) rights in 1995.

The Article 27.3 (b) of TRIPs agreement makes it mandatory for the member countries to provide protection for plant varieties either by patents, by an effective *sui-generis* system, or any combination thereof for effective protection of intellectual property. The Government of India enacted a Protection of Plant Varieties and Farmers Rights Act (PPVFRA) during 2001, and consequently a National Plant Authority is established to facilitate the registration of plant varieties. The guidelines of PPVFRA are used for characterization and registration of newly bred cultivars. Hence, this book attempts the detailed characterization of sweet sorghum cultivars and female hybrid parents following the guidelines of PPVFRA in order to help the researchers, entrepreneurs, farmers and other stakeholders to identify the available sweet sorghum cultivars and understand their yield potential in tropics.

References

Almodares A, Hadi MR (2009) Production of bio-ethanol from sweet sorghum: a review. Afr J Agric Res 4:772–780

Audilakshmi S, Aruna C, Garud TB, Nayakar NY, Atale SB, Veerabadhiran P, Dayakar B Rao, Ratnavathi CV, Indira S (2005) A technique to enhance the quality and market value of rainy season sorghum grain. Crop Prot 24:251–258

Cosentino SL, Mantineo M, Testa G (2012) Water and nitrogen balance of sweet sorghum (*Sorghum bicolor* Moench (L.)) cv. Keller under semi-arid conditions. Ind Crops Prod 36:329–342

de Wet JMJ (1978) Systematics and evolution of *Sorghum* sect. *Sorghum* (Gramineae). Am J Bot 65:477–484

de Wet JMJ, Harlan JR (1971) The origin and domestication of *Sorghum bicolor*. Econ Bot 25:128–135

Doggett H (1976) Sorghum: *Sorghum bicolor* (Gramineae-Andropogoneae). In: Si'mmonds NW (ed) Evolution of crop plants. Longman, London, pp 112–117

Doggett H (1988) Sorghum. Longman Scientific and Technical, London

Duvall MR, Doebley JF (1990) Restriction site variation in the chloroplast genome of *Sorghum* (Poaceae). Syst Bot 15:472–480

FAO (1995) Sorghum and millets in human nutrition. Chapter 1: Introduction. FAO Food and Nutrition Series, No. 27, ISBN 92-5-103381-1. FAO Corporate Document Repository. http://www.fao.org/docrep/T0818e/T0818E00.htm#Contents. Accessed 20 March 2012

Garber ED (1950) Cytotaxonomic studies in the genus *Sorghum*. Univ Calif Publ Bot 23:283–361

Girase JR (2010) Evaluation of the economic feasibility of grain sorghum, sweet sorghum, and switch grass as alternative feedstocks for ethanol production in the Texas panhandle. MS thesis, West Texas A&M University, Canyon, Texas, USA, p 101

Gnansounou E, Dauriat A, Wyman CE (2005) Refining sweet sorghum to ethanol and sugar: economic trade-offs in the context of North China. Bioresour Technol 96:985–1002

Harlan JR, de Wet JMJ (1972) A simplified classification of cultivated sorghums. Crop Sci 12:172–176

House LR (1980) A guide to sorghum breeding. ICRISAT, Patancheru, p 238

Kumar CG, Fatima A, Srinivasa Rao P, Reddy BVS, Rathore A, Rao RN, Khalid S, Ashok Kumar A, Kamal A (2010) Characterization of improved sweet sorghum genotypes for biochemical parameters, sugar yield and its attributes at different phenological stages. Sugar Tech 12:322–328

Lazarides M, Hacker JB, Andrew MH (1991) Taxonomy, cytology and ecology of indigenous Australian Sorghums (*Sorghum* Moench: Andropogoneae: Poaceae). Aust Syst Bot 4:591–635

Maiti R (1996) Sorghum science. Oxford and IBH Publishing, New Delhi, p 352

Purseglove JW (1972) Tropical crops: monocotyledons, vol 1. Longman Group Limited, London, p 334

Quinby JR, Karper RE (1947) The effect of short photoperiod on sorghum varieties and first generation hybrids. J Agric Res 75:295–300

Reddy BVS, Ramesh S, Ashok Kumar A, Wani SP, Ortiz R, Ceballos H, Sreedevi TK (2008) Bio-fuel crops research for energy security and rural development in developing countries. Bioenergy Res 1:248–258

Reddy BVS, Ramesh S, Sanjana P Reddy, Ramaiah B, Salimath PM, Kachapur R (2005) Sweet sorghum—a potential alternative raw material for bioethanol and bio-energy. Int Sorghum Millets Newsl 46:79–86

Sanderson MA, Jones RM, Ward J, Wolfe R (1992) Silage sorghum performance trial at Stephen-ville. Forage Research in Texas. Report PR-5018. Texas Agricultural Experimental Station, Stephenville, USA

Spangler RE (2003) Taxonomy of *Sarga*, *Sorghum* and *Vacoparis* (Poaceae: Andropogoneae). Aust Syst Bot 16:279–299

Srinivasa Rao P, Rao SS, Seetharama N, Umakanth AV, Sanjana Reddy P, Reddy BVS, Gowda CLL (2009) Sweet sorghum for biofuel and strategies for its improvement. International Crops Research Institute for the Semi-Arid Tropics, Patancheru. Information Bulletin No. 77. 80 pages. ISBN 978-92-9066-518-2

Srinivasa Rao P, Reddy BVS, Blümmel M, Subbarao GV, Chandraraj K, Sanjana Reddy P, Parathasarathy Rao P (2010) Sweet sorghum as a biofuel feedstock: can there be food-feed-fuel trade-offs? http://www.wgcrop.icidonline.org/sweet%20sorghumdec09.pdf. Accessed 23 March 2012)

Srinivasa Rao P, Sanjana Reddy P, Rathore A, Reddy BVS, Panwar S (2011a) Application of GGE biplot and AMMI model to evaluate sweet sorghum hybrids for genotype × environment interaction and seasonal adaptation. Indian J Agric Sci 81:438–444

Srinivasa Rao P, Jayalakshmi M, Kumar CG, Kamal A, Reddy BVS (2011b) Response of fertilizer treatment on agronomic and biochemical traits in main and ratoon crops of sweet sorghum (*Sorghum bicolor* (L) Moench) cultivar ICSV 93046. Balancing Sugar and Energy Production in Developing Countries: Sustainable Technologies and Marketing Strategies. In: Proceedings of the 4th IAPSIT international sugar conference IS-2011. Hotel Parkland Exotica, New Delhi, pp 160–164, 21–25 November

USDA, ARS (2006) National genetic resources program. Germplasm resources information network (GRIN) [online database]. National Germplasm Resources Laboratory, Beltsville. http://www.ars-grin.gov/cgi-bin/npgs/html

Vinall HN, Stephens JC, Martin JH (1936) Identification, history, and distribution of common sorghum varieties. Technical Bulletin No. 506, US Department of Agriculture, Washington, DC

Wortmann CS, Liska AJ, Ferguson RB, Lyon DJ, Klein RM, Dweikat I (2010) Dryland performance of sweet sorghum and grain crops for biofuel. Agron J 102:319–326

Ziggers D (2006) Sorghum—the multipurposes grass. Feed Tech 5:20–23

Methodology, Results and Discussion

C. Ganesh Kumar and P. Srinivasa Rao

Abstract There are many research and review articles published on sweet sorghum. However, no single publication gives a detailed account of the morpho-biochemical traits of improved tropical sweet sorghum cultivars. This chapter gives detailed account of the materials used, methods followed for data collection and analysis to characterise sweet sorghum genotypes following the guidelines of Protection of Plant Varieties and Farmers Rights Act, 2001 (PPVFRA). The pooled analysis of variance for quantitative traits revealed that these cultivars had significant differences between them for the expression of all the quantitative characters under study for both the seasons. The results revealed that the productivity levels of tropical sweet sorghums during post-rainy season (October–March) are generally low due to photo-sensitivity and thermo-sensitivity of the genotypes *vis-a-vis* that of rainy season (June–October) and necessitates identifying new sources/alleles contributing to both biomass and sugar yield.

Keywords Methodology · Discussion · Restorers · Varieties · Hybrids · Breeding · Quantitative traits · Glucose · Fructose · Sucrose · Brix%

The chapter is discussed under three sections, i.e. materials, data collection, result and discussion.

C. Ganesh Kumar
Chemical Biology Laboratory, CSIR-Indian Institute of Chemical Technology (CSIR-IICT), Uppal Road, Hyderabad 500607, India

P. Srinivasa Rao (✉)
International Crops Research Institute for the Semi-Arid Tropics (ICRISAT), Patancheru 502324, India
e-mail: psrao72@gmail.com

P. S. Rao and C. G. Kumar (eds.), *Characterization of Improved Sweet Sorghum Cultivars*, SpringerBriefs in Agriculture, DOI: 10.1007/978-81-322-0783-2_2, © The Author(s) 2013

1 Materials

Eleven established restorers/varieties (ICSV 700, ICSV 25272, ICSV 25274, ICSV 25275, ICSV 25280, ICSV 93046, SPV 422, SSV 74, SSV 84 and CSV 24SS), six female hybrid parents (ICSB 38, ICSB 474, ICSB 675, ICSB 702, ICSB 724 and ICSB 731) and six hybrids adapted to rainy season (ICSSH 25, ICSSH 28, ICSSH 29, ICSSH 30, ICSSH 31, ICSSH 39, ICSSH 58 and CSH 22SS) and seven established restorers/varieties (ICSV 700, ICSV 25279, ICSV 25284, ICSV 93046, SSV 74, SSV 84 and CSV 24SS), five female hybrid parents (ICSB 38, ICSB 474, ICSB 502, ICSB 675 and ICSB 731) and five hybrids (ICSSH 25, ICSSH 28, ICSSH 58, ICSSH 76 and CSH 22SS) adapted to post-rainy season bred at ICRISAT, Directorate of Sorghum Research (DSR), University of Agricultural Sciences (UAS) Dharwad and Mahatma Phule Krishi Vidyapeeth (MPKV), were evaluated during 2010 rainy season (June–October) and 2010–2011 post-rainy season (October–March) in vertisols (deep black soils) at the research farm, ICRISAT, Patancheru, India. The experimental site is located at an altitude of 545 m above mean sea level, latitude of 17.53°N and latitude of 78.27°E. The site receives an average annual rainfall of 897 mm (average of 32 years from 1974 to 2005). The entries were planted in four rows, 4 m long, with a row spacing of 0.60 and 0.15 m between the plants within a row, following a randomized complete block design (RCBD) in three replications. The recommended crop production and protection packages were followed to raise a healthy crop.

2 Data Collection

The observations on twenty two quantitative traits were taken on 10 random plants in each plot for plant height (m), panicle length (cm), stalk diameter (cm), leaf blade length (cm), leaf blade width (cm), exsersion length (cm), time to panicle emergence (days), panicle length (cm), panicle primary branch length (cm), glume coverage (%), stalk yield (t ha^{-1}), juice yield (t ha^{-1}), total soluble solids (TSS) or Brix (%), sugar yield (t ha^{-1}), seed restoration (%) and 1000-grain weight (g). The seed restoration (%) was collected only in hybrids as varieties are self-compatible (Srinivasarao et al. 2009; Wortmann et al. 2010). The restoration problems in sorghum hybrids arises due to incompatible reaction male sterile cytoplasm with fertility restorer genes of restorers, which is common in A$_2$, A$_3$ and A$_4$ male sterile cytoplasms. The days to 50 % flowering was recorded on plot basis when the main panicles of 50 % of the plants in the plots had full stigma emergence. The sugar concentration in the stems was estimated in terms of Brix% using a hand-held pocket refractometer (Model PAL, Atago Co. Ltd., Tokyo, Japan) based on the extracted juice samples taken from each plot. The pH was recorded using a microprocessor-based pH meter (Model DPH506, Global Electronics, Hyderabad, India). The electrical conductivity (EC) probe is dipped in the juice sample and measurements were done using a microprocessor-based EC analyzer

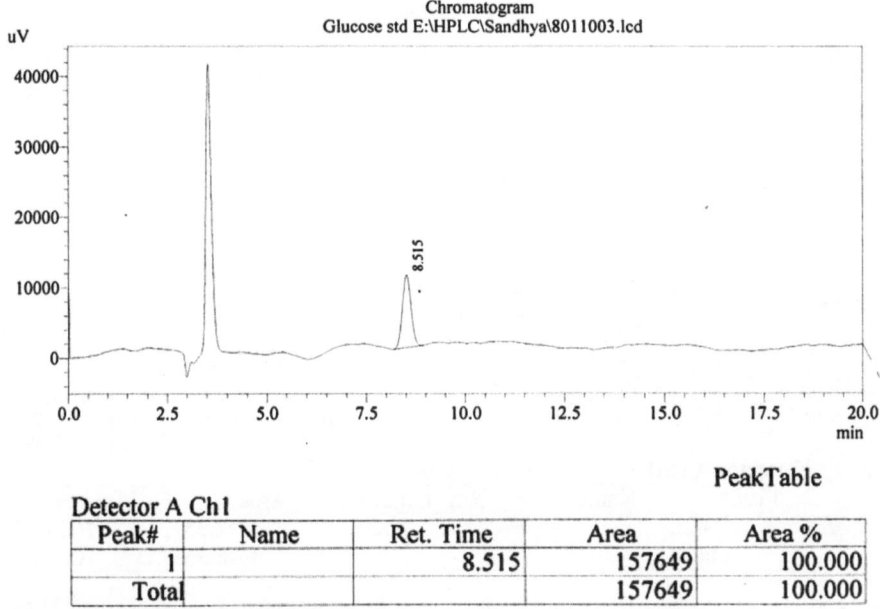

Fig. 1 Standard chromatogram of glucose

Detector A Ch1

PeakTable

Peak#	Name	Ret. Time	Area	Area %
1		8.515	157649	100.000
Total			157649	100.000

(Model CM 180, Elico Limited, Hyderabad, India). Between two different sample readings, the refractometer, pH and EC probes were cleaned with distilled water and dried with a paper towel. Sugar profiling to determine the relative percentages of hexose sugars like glucose, fructose and sucrose present in the sweet sorghum juice of each genotype were analyzed on a HPLC system (Shimadzu, Kyoto, Japan) equipped with a Lichro CART 250-4,6 Lichrospher 100 NH$_2$ (5 μm, Merck KGaA, Darmstadt, Germany). The detection of the separated sugars was carried out with a refractive index detector (Model RID-10A, Shimadzu, Kyoto, Japan) using a mobile phase of acetonitrile–water (78:22, v/v) at a flow rate of 2.0 ml min^{-1} under isocratic mode and the column temperature was maintained at 40 °C. All solvents for mobile phase optimization were degassed before use. Standard stock solution (1,000 μg ml^{-1}) of different sugars was prepared in Milli-Q distilled water as a diluent for calibrating the HPLC system. The juice sample analysis was carried out by manual injection of 20 μl of pre-filtered sample. The data acquisition and analysis was carried out using LC solutions software (version 1.24 SP2) (Shimadzu, Kyoto, Japan). The concentration of each sugar in the juice was determined using peak area from the chromatograms and expressed in terms of percentage of total sugars (Kumar et al. 2010). The standard chromatograms for glucose, fructose and sucrose are given in Figs. 1, 2 and 3.

Data were also recorded on 16 other traits which included anthocyanin colouration of coleoptile, anthocyanin colouration of leaf sheath, leaf midrib colour, flag leaf midrib colouration, presence/absence of awns on lemma, anthocyanin

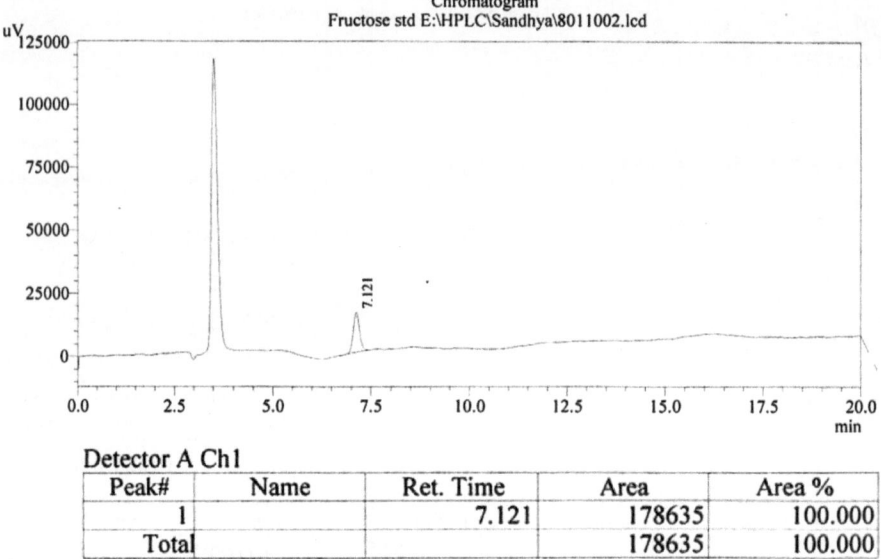

Fig. 2 Standard chromatogram of fructose

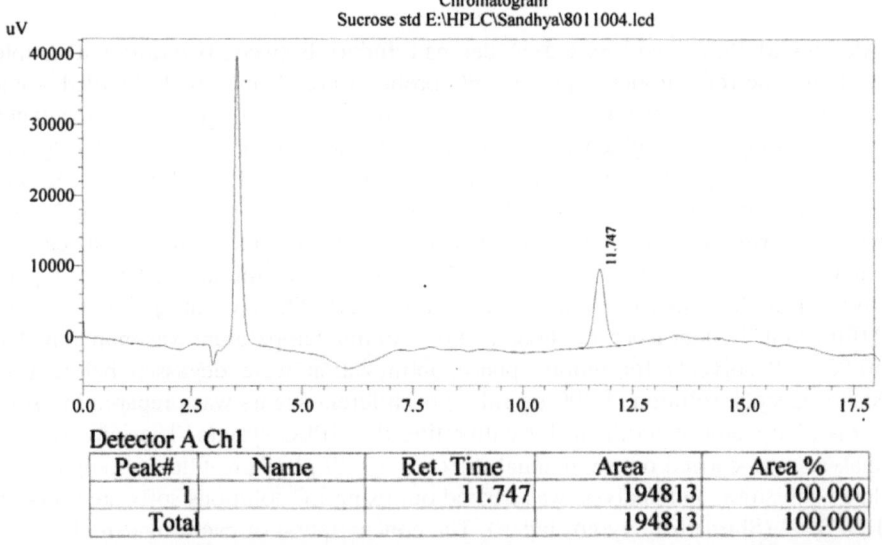

Fig. 3 Standard chromatogram of sucrose

colouration of stigma, yellow colouration of stigma, length of stigma, length of floret, anther length, color of anther, glume color, panicle density, panicle shape, panicle neck length, threshability, caryopsis (grain) color, grain shape-profile view and dorsal view, germ size on the grain, endosperm texture, albumen color and grain lustre for which data were recorded (Reddy et al. 2006) on the basis of visual assessment of individual plants (or parts of plants) within a plot; or it was based on visual assessment of group of plants (or parts of plants) in a plot for traits such as plant growth habit, panicle shape, panicle density, grain color and grain shape. The mean plot values of the quantitative traits measured were subjected to analysis of variance (ANOVA) for each season using Genstat 14.1 software. The data analysis was done separately for varieties and hybrids as one group and treating all female hybrid parents (B-lines) as another group owing to their distinct genetic differences.

3 Results and Discussion

In sweet sorghum improvement program at ICRISAT, hybrid parental lines (A/B pairs) and varieties are developed with considerable morphological diversity and then designated based on agronomic performance and resistance to shootfly. The tropical sweet sorghums are photo- and thermo-sensitive and flowers when the day length is less than 12 h 15 min. The biomass yield and sugar yield are high during rainy season as compared to that of post-rainy season. The extent of variation attributable to seasonal effect on these genotypes may vary depending on the environments (Srinivasarao et al. 2011). However, pooled analysis of variance for quantitative traits revealed that these cultivars had significant differences between them for the expression of all the quantitative characters under study (Table 1). Further the magnitude of variation is highly influenced by the environment, particularly for sugar yield and related traits (Srinivasarao et al. 2011). The mean performance of sweet sorghum varieties and hybrids adapted to rainy season are presented in Table 2. The mean of key sugar related traits are plant height: 3.39 m (range: 3–3.8 m), days to 50 % flowering: 80 days (range: 72–87 days), stalk yield: 47.9 t ha^{-1} (range: 36.6–60.6 t ha^{-1}), juice yield:18.9 t ha^{-1} (range: 13.5–29.4 t ha^{-1}), Brix%: 17.3 (range: 16–20) and sugar yield: 2.6 t ha^{-1} (range: 1.8–3.5 t ha^{-1}). The mean of the key biochemical parameters are pH: 5.0 (range: 4.0–5.0), electrical conductivity: 8.6 mS cm^{-1} (range: 7.1–10.3 mS cm^{-1}), fructose: 0.8 % (range: 0.2–2.83 %), glucose: 0.96 % (range: 0.23–1.86 %) and sucrose: 6.29 % (range: 2.63–9.90 %). In general varieties have recorded 5.6 % more Brix% while hybrids are taller by 0.2 m and flower 5.4 days earlier. The hybrids have recorded 12.8 % higher stalk yield, 24.4 % more juice yield and 17 % higher sugar yield *vis-a-vis* varieties in rainy season. Hence, the available heterosis for traits like plant height, stalk yield and juice yield needs to be exploited favorably to develop high sugar yielding hybrids (Srinivasarao et al. 2009, 2010; Reddy et al. 2011; Kumar et al. 2011). As the distillery needs to be operated for longer period of the year to be economically viable, the earliness in rainy season hybrids cannot be ignored to develop hybrids of choice with different maturity groups.

The mean performance of sweet sorghum female hybrid parents (6) adapted to rainy season are presented in Table 3. The mean of key sugar related traits are

C. Ganesh Kumar and P. Srinivasa Rao

Table 1 Mean squares for quantitative traits for sweet sorghum varieties and hybrids during 2010 rainy season

Source of variation	d.f	Plant height (m)	Stem diameter (mm)	Leaf length (cm)	Leaf width (cm)	Exersion length (cm)	Panicle branch length (cm)	Panicle length (cm)	Glume coverage (%)	1,000 grain weight (g)	Days to 50 % flowering
Replication	2	0.02	12.42	142.47	15.52	16.54	1.02	26.58	8.75	0.25	0.8
Genotype	19	0.03*	10.37**	130.21*	3.97*	86.80**	10.98**	34.83**	840.43**	179.59**	79.01**
Residual	38	0.02	4.2	63.52	2.06	9.32	1.42	6.54	3.93	0.05	2.31

Source of variation	Stalk yield (t ha^{-1})	Juice yield (t ha^{-1})	Juice extraction (%)	Brix (%)	Grain yield (t ha^{-1})	Seed restoration (%)	pH	EC (ms/cm)	Fructose (%)	Glucose (%)	Sucrose (%)
Replication	0.2	0.18	16.09	4.77	0.11	71.64	0.8	0.47	2.77	5.04	7.07
Genotype	1651.22**	329.55**	88.09**	17.90**	2.92**	117.23**	0.19*	2.49**	1.60**	0.82*	12.10**
Residual	0.45	0.12	1.4	0.86	0.01	2.26	0.09	0.29	0.45	0.34	3.38

d.f. degrees of freedom; *significant at P < 0.05; **significant at P < 0.01

Table 2 Mean performance of sweet sorghum varieties and hybrids for quantitative traits during 2010 rainy season

S.No	Genotype	Plant height (m)	Stem diameter (mm)	Leaf length (cm)	Leaf width (cm)	Exersion length (cm)	Panicle branch length (cm)	Panicle length (cm)	Glume coverage (%)	1,000 grain weight (g)	Days to 50% flowering	Stalk yield (t ha^{-1})
1	SPV 422	3.1	21.60	88	7	4	7	20	25	35	86	47.58
2	ICSV 25274	3.5	19.65	90	11	8	7	23	23	49	87	49.08
3	ICSV 25280	3.3	15.11	71	8	6	6	17	50	34	83	45.41
4	ICSV 25275	3.3	15.20	70	8	7	5	16	50	30	81	46.57
5	ICSV 25272	3.1	14.50	81	7	14	9	23	47	33	74	55.57
6	RSSV 9	3.3	17.75	75	10	3	5	18	47	26	85	58.87
7	ICSV 700	3.4	17.62	73	8	5	6	20	50	28	84	45.63
8	SSV 84	3.2	18.34	88	11	7	7	21	50	25	87	38.74
9	SSV 74	3.3	18.21	76	9	7	8	19	25	36	84	41.54
10	ICSV 93046	3.6	17.89	74	8	5	7	18	50	27	86	53.96
11	ICSSH 31	3.1	14.16	85	8	12	8	28	75	32	75	51.16
12	ICSSH 25	3.5	18.05	77	9	8	8	24	25	28	77	54.77
13	ICSSH 39	3.3	16.37	82	10	10	10	26	25	36	74	46.90
14	ICSSH 30	3.5	18.78	82	10	13	8	24	25	44	77	52.02
15	ICSSH 29	3.8	17.76	91	10	9	7	25	25	27	82	58.49
16	ICSSH 58	3.7	14.97	88	9	9	7	21	50	25	79	57.50
17	ICSSH 28	3.8	17.47	80	10	12	7	25	25	47	76	63.64
18	CSH 22 SS	3.3	18.54	86	9	19	8	25	25	30	84	51.27
	Mean	3.4	17.26	81	9	10	8	22	40	33	80	46.69
	Minimum	3.1	14.16	70	7	3	5	16	23	25	74	38.74
	Maximum	3.8	21.60	91	11	19	10	28	75	49	87	63.64
	LSD (5 % Level)	0.24	3.39	13.17	2.37	5.05	1.97	4.23	3.28	0.37	2.51	1.10
	CV (%)	1.00	4.60	3.30	10.00	9.20	3.00	5.20	1.70	0.30	0.20	0.20

(continued)

Table 2 (continued)

S.No	Genotype	Juice yield (t ha⁻¹)	Juice extraction (%)	Brix (%)	Sugar yield (t ha⁻¹)	Grain yield (t ha⁻¹)	Seed restoration (%)	pH	EC (ms/cm)	Fructose (%)	Glucose (%)	Sucrose (%)
1	SPV 422	18.71	39	20	2.8	3.1	*	5	8.8	0.23	0.55	3.59
2	ICSV 25274	19.59	40	20	2.9	2.6	*	5	8.1	0.44	0.96	2.63
3	ICSV 25280	24.35	54	17	3.1	2.4	*	5	7.6	0.42	1.86	4.91
4	ICSV 25275	23.51	50	16	2.8	2.7	*	5	7.7	0.44	1.72	7.73
5	ICSV 25272	26.34	47	16	3.2	3.4	*	5	9.5	1.09	0.81	6.4
6	RSSV 9	21.43	36	16	2.6	1.5	*	4	9.4	2.83	1.62	4.68
7	ICSV 700	17.4	38	16	2.1	3.1	*	5	8.4	0.21	0.71	4.94
8	SSV 84	13.53	35	19	1.9	3.3	*	5	7.1	0.38	0.69	3.71
9	SSV 74	14.86	36	16	1.8	1.5	*	5	7.8	0.32	1.49	4.99
10	ICSV 93046	25.58	47	16	3.1	2.7	*	5	8.9	0.2	0.65	4.01
11	ICSSH 31	22.36	44	16	2.7	3.3	91	5	9.3	1.38	0.23	8.41
12	ICSSH 25	28.45	52	17	3.6	3.1	86	5	10.3	0.29	1.28	6.25
13	ICSSH 39	22.45	48	18	3	2.2	83	5	9.5	0.37	1.1	8.07
14	ICSSH 30	24.5	47	18	3.3	2.2	92	5	7.5	2.01	0.56	7.77
15	ICSSH 29	25.44	43	19	3.6	2.8	82	5	7.5	1.53	0.37	7.61
16	ICSSH 58	29.43	51	17	3.8	2.9	93	5	8.2	1.19	0.34	9.9
17	ICSSH 28	28.34	45	16	3.4	2.4	82	5	9.6	0.54	1.52	6.32
18	CSH 22 SS	23.38	46	19	3.3	3.5	77	5	8.9	1.57	0.3	8.48
	Mean	18.85	33	17.3	2.9	2.7	85	5	8.6	0.8	0.96	6.29
	Minimum	13.53	35	16	1.8	1.5	77	4	7.1	0.2	0.23	2.63
	Maximum	29.43	54	20	3.8	3.5	93	5	10.3	2.83	1.86	9.9
	LSD (5 % Level)	0.57	1.96	1.53	0.3	0.16	2.6	0.49	0.89	1.11	0.97	3.04
	CV (%)	0.5	2.7	3	7.9	3.1	2.2	4	1.8	16.3	12.4	9.5

LSD least significant difference; *CV* coefficient of variation

Table 3 Mean performance of sweet sorghum female hybrid parental lines for quantitative traits during 2010 rainy season

S.No	Genotype	Plant height (m)	Stem diameter in (mm)	Leaf length (cm)	Leaf width (cm)	Exertion length (cm)	Panicle branch length (cm)	Panicle length (cm)	Glume coverage (%)	1,000 grain weight (g)	Days to 50 % flowering	Stalk yield (t ha^{-1})
1	ICSB 38	1.4	17.67	81	8	24	8	32	25	26	74	16.0
2	ICSB 474	2.1	14.1	71	8	10	6	21	50	43	70	23.1
3	ICSB 675	1.2	15.02	86	9	0	0	21	25	16	76	21.4
4	ICSB 702	1.6	15	79	7	16	7	23	25	37	67	17.4
5	ICSB 724	1.9	14.98	76	8	9	7	20	25	27	76	24.8
6	ICSB 731	2	16.38	82	8	11	6	22	25	29	74	20.3
	Mean	1.9	15.53	79	8	12	6	23	29	30	73	20.5
	Minimum	1.2	14.1	71	7	0	0	20	25	16	67	16.0
	Maximum	2.1	17.67	86	9	24	8	32	50	43	76	24.8
	LSD (5 % Level)	0.22	4.6	10.65	2.34	3.23	1.94	3.78	*	*	2.73	0.72
	CV (%)	1.8	9.6	1.7	19.3	9.7	13	9.6	0	0	0.9	0.7

S.No	Genotype	Juice yield (t ha^{-1})	Juice extraction (%)	Brix (%)	Sugar yield (t ha^{-1})	Grain yield (t ha^{-1})	pH	EC (ms/cm)	Fructose (%)	Glucose (%)	Sucrose (%)
1	ICSB 38	4.4	27.5	9	0.3	2.3	5.5	14.7	0.65	2.07	1.71
2	ICSB 474	8	34.4	11	0.66	1.4	5.1	10.2	0.28	1.17	5.67
3	ICSB 675	6.7	31.3	12	0.6	0.6	5.3	12.2	0.72	2.39	1.74
4	ICSB 702	7.6	43.6	13	0.74	4.1	5.3	11.5	0.2	1.02	4.83
5	ICSB 724	9.7	39.1	13	0.94	4.3	5.3	9.7	0.37	1.41	4.62
6	ICSB 731	9.2	45	14	0.96	3.9	5.1	9.1	0.34	0.86	8.87
	Mean	7.6	36.8	12	0.7	2.8	5.2	11.2	0.43	1.49	4.57
	Minimum	5.4	27.5	9	0.3	0.6	5.1	9.1	0.2	0.86	1.71
	Maximum	9.7	45	14	0.96	4.3	5.5	14.7	0.72	2.39	8.87
	LSD (5 % Level)	0.49	1.4	1.09	0.1	0.43	0.17	1.66	0.27	0.78	2.3
	CV (%)	1.6	3.2	5.8	11.4	0.4	1	5.6	39.7	36.7	33.9

LSD least significant difference; CV coefficient of variation plant height: 1.7 m (range: 1.2–2.1 m), days to 50 % flowering: 73 days (range: 67–76 days), stalk yield: 20.5 t ha^{-1} (range: 16–24.8 t ha^{-1}), juice yield:7.6 - t ha^{-1} (range: 5.4–9.7 t ha^{-1}), Brix%: 12 (range: 9–14) and sugar yield: 0.7 t ha^{-1} (range: 0.3–0.96 t ha^{-1}). The mean of the key biochemical parameters are pH: 5.2 (range: 5.1–5.5), electrical conductivity: 11.2 mS cm^{-1} (range: 9.2–14.7 mS cm^{-1}), fructose: 0.43 % (range: 0.2–0.72 %), glucose: 1.49 % (range: 0.86–2.39 %) and sucrose: 4.57 % (range: 1.71–8.87 %). The high sugar yielding B-lines such as ICSB 474, ICSB 702, ICSB 724 and ICSB 731 can be utilized in the breeding programmes to develop highly productive hybrids for ethanol production. The detailed characteristics of sweet sorghum cultivars and female hybrid parents adapted to rainy season were discussed in Chap. 3.

The productivity levels of tropical sweet sorghums during post-rainy season (October–March) are generally low due to photo-sensitivity and thermo-sensitivity of the genotypes (Srinivasarao et al. 2009; Kumar et al. 2010). As the sugar accumulation is a function of diurnal and nocturnal temperature differences besides complex genotype $\times$ environment (G $\times$ E) interactions. The pooled analysis of variance for quantitative traits revealed significant differences among the cultivars studied (data not shown). The mean performance of sweet sorghum varieties and hybrids adapted to post-rainy season are presented in Table 4. The cultivar mean for sugar related traits are plant height: 1.87 m (range: 1.5–2.3 m), days to 50 % flowering: 78 days (range: 69–83 days), stalk yield: 29.9 t ha^{-1} (range: 20.1–38.1 t ha^{-1}), juice yield:10.9 t ha^{-1} (range: 6.9–15.8 t ha^{-1}), Brix%: 14 (range: 10–17) and sugar yield: 1.16 t ha^{-1} (range: 0.67–2.02 t ha^{-1}). The mean of the key biochemical parameters are pH: 5.3 (range: 5.2–5.4), electrical conductivity: 13.66 mS cm^{-1} (range: 9.6–18.0 mS cm^{-1}), fructose: 1.33 % (range: 0.95–2.04 %), glucose: 1.11 % (range: 0.79–1.47 %) and sucrose: 4.5 % (range: 2.38–7.35 %).

The post-rainy season adapted hybrids exhibited 30.1 % superiority for stalk yield, 59 % higher sugar yield besides 10 % higher grain yield over that of varieties. The Brix% levels are same in both the groups; however there is a reduction of total soluble solids by 25 % in post-rainy season as compared to that of rainy season. In case of sugar yield across two seasons, the average post-rainy season sugar productivity (1.16 t ha^{-1}) is 156 % which is lower that of the rainy season level (2.98 t ha^{-1}). This data further reiterates that the genetic pool of sweet sorghums needs to be broadened by attempting novel approaches of either creating variability (Targeting Induced Local Lesions IN Genomes-TILLING) or introgression of novel alleles by wide hybridization etc. (McCallum et al. 2000; Rooney et al. 2007).

The performance of female hybrid parents adapted to post-rainy season is presented in Table 5. The mean values for sugar related traits of B-lines are plant height: 1.02 m (range: 0.8–1.5 m), days to 50 % flowering: 78 days (range: 74–76 days), stalk yield: 14.6 t ha^{-1} (range: 9.6–17 t ha^{-1}), juice yield: 4.42 t ha^{-1} (range: 2.8–5.9 t ha^{-1}), Brix%: 10.8 (range: 9–13) and sugar yield: 0.35 t ha^{-1} (range: 0.27–0.44 t ha^{-1}). The mean of the key biochemical param-eters are pH: 5.3 (range: 5.2–5.4), electrical conductivity: 16.66 mS cm^{-1} (range: 13.47–18.77 mS cm^{-1}), fructose: 1.55 % (range: 1.04–2.23 %), glucose: 1.07 %

Table 4 Mean performance of sweet sorghum varieties and hybrids for quantitative traits during 2010 post-rainy season

S.No	Genotype	Plant height (m)	Stem diameter (mm)	Leaf length (cm)	Leaf width (cm)	Exersion length (cm)	Panicle branch length (cm)	Panicle length (cm)	Glume coverage (%)	1000 grain weight (g)	Days to 50 % flowering	Stalk yield (t ha⁻¹)
1	RSSV 9	1.9	14.43	69	7	6	5	13	33	33	69	27.5
2	ICSV 25272	2	15.43	67	8	14	5	16	48	33	83	25.8
3	ICSV 25274	1.8	14.08	67	8	13	7	18	25	41	72	29.7
4	SSV 84	1.5	13.65	77	8	11	6	17	25	32	72	20.9
5	SSV 74	1.7	12.77	63	7	11	8	18	25	43	73	23.2
6	ICSV 700	2	13.22	54	8	8	6	17	67	30	81	24.7
7	ICSV 93046	2	15.29	51	7	11	6	16	42	33	83	33.9
8	ICSSH 28	2.1	12.06	65	8	17	8	23	25	36	73	31.7
9	ICSSH 58	2.3	14.55	64	8	15	7	17	25	26	89	38.1
10	ICSSH 25	1.9	14.39	66	7	21	9	22	33	32	77	31.7
11	CSH 22SS	2.1	13.81	63	8	27	8	23	25	25	73	36.7
12	ICSSH 76	1.6	13.61	83	8	22	8	25	33	31	93	34.4
	Mean	1.87	13.94	66	8	15	7	19	34	33	78	29.9
	Minimum	1.5	12.06	51	7	6	5	13	25	25	69	20.9
	Maximum	2.3	15.43	83	8	27	9	25	67	43	93	38.1
	LSD (5 % Level)	0.15	4.44	13.24	1.64	7.76	1.42	2.93	21.15	4.14	6.44	3.49
	CV (%)	6.1	3.6	3	2.6	6.2	3.9	3.4	7.8	1	0.6	8.6

S.No	Genotype	Juice yield (t ha⁻¹)	Juice extraction (%)	Brix (%)	Sugar yield (t ha⁻¹)	Grain yield (t ha⁻¹)	Seed restoration (%)	pH	EC (ms/cm)	Fructose (%)	Glucose (%)	Sucrose (%)
1	RSSV 9	10.6	38.7	14	1.12	3.1	93	5.2	16	1.17	1.45	4.08
2	ICSV 25272	9.1	35.1	16	1.09	3.5	93	5.3	12.9	1.15	0.79	7.35
3	ICSV 25274	7.7	25.7	14	0.8	2.7	93	5.3	14.87	1.67	1.33	3.88
4	SSV 84	9.1	43.4	10	0.68	3	93	5.2	16.6	1.06	1.15	2.71

(continued)

Table 4 (continued)

S.No	Genotype	Juice yield (t ha^{-1})	Juice extraction (%)	Brix (%)	Sugar yield (t ha^{-1})	Grain yield (t ha^{-1})	Seed restoration (%)	pH	EC (ms/cm)	Fructose (%)	Glucose (%)	Sucrose (%)
5	SSV 74	8.4	36	13	0.82	2.7	93	5.2	12.87	1.14	1.47	5.14
6	ICSV 700	6.9	27.7	13	0.67	2.6	93	5.3	11.27	2.04	1	4.93
7	ICSV 93046	11.2	32.9	16	1.34	3.1	93	5.3	10.53	1.71	0.97	5.59
8	ICSSH 28	11.9	37.5	15	1.34	2.6	91	5.2	15.43	1.05	0.95	3.92
9	ICSSH 58	15.8	41.5	17	2.02	3.2	85	5.2	9.6	1.49	1.09	6.87
10	ICSSH 25	13.3	41.9	14	1.39	2.7	94	5.2	13.03	1.23	1.26	3.84
11	CSH 22SS	13.1	35.6	13	1.27	3.7	100	5.3	18	0.95	0.79	2.38
12	ICSSH 76	14.2	41.4	13	1.39	4.2	97	5.4	12.87	1.29	1.07	3.34
	Mean	10.9	36.4	14	1.16	3.09	93	5.3	13.66	1.33	1.11	4.5
	Minimum	6.9	27.7	10	0.67	2.6	85	5.2	9.6	0.95	0.79	2.38
	Maximum	15.8	43.4	17	2.02	4.2	100	5.4	18	2.04	1.47	7.35
	LSD (5 % Level)	4.46	5.05	2.92	0.3	1.03	6.19	0.12	3.35	0.66	0.37	2.44
	CV (%)	13.7	8	4.1	8.6	3.2	1.6	0.1	2.9	8.2	6	12.4

LSD least significant difference; *CV* coefficient of variation

Table 5 Mean performance of sweet sorghum female hybrid parental lines for quantitative traits during 2010 post-rainy season

S.No	Genotype	Plant height (m)	Stem diameter (mm)	Leaf length (cm)	Leaf width (cm)	Exersion length (cm)	Panicle branch length (cm)	Panicle length (cm)	Glume coverage (%)	1,000 grain weight (g)	Days to 50 % flowering	Stalk yield (t ha⁻¹)
1	ICSB 38	0.9	15.98	71	7	20	9	27	25	28	75	9.6
2	ICSB 474	1	12.58	70	7	12	7	24	67	27	74	17
3	ICSB 502	0.9	16.4	75	7	6	6	25	25	31	78	15.7
4	ICSB 731	1.5	16.38	62	7	12	6	18	25	31	77	17
5	ICSB 675	0.8	19.69	73	8	7	9	27	25	35	76	13.7
	Mean	1	16.21	70	7	11	7	24	33	31	76	10.2
	Minimum	0.8	12.58	62	7	6	6	18	25	27	74	9.6
	Maximum	1.5	19.69	75	8	20	9	27	67	35	78	17
	LSD (5 % Level)	0.05	7.2	15.62	1.79	13.32	1.72	4.7	12.15	4.16	2.41	4
	CV (%)	1.2	9	2	5.2	28.7	2.9	5.1	8.7	3.5	0.8	21.8

S.No	Genotype	Juice yield (t ha⁻¹)	Juice extraction (%)	Brix (%)	Sugar yield (t ha⁻¹)	Grain yield (t ha⁻¹)	pH	EC (ms/cm)	Fructose (%)	Glucose (%)	Sucrose (%)
1	ICSB 38	3.3	34.1	11	0.27	3.8	5.3	16.87	1.47	0.96	3.04
2	ICSB 474	4.2	24.4	11	0.34	3.3	5.2	17.37	1.04	1.19	2.68
3	ICSB 502	3.9	24.8	13	0.38	2.5	5.4	13.47	2.23	0.98	4.11
4	ICSB 731	5.9	34.8	10	0.44	2.5	5.3	16.83	1.45	1.2	2.64
5	ICSB 675	4.8	34.8	9	0.32	3.5	5.3	18.77	1.54	1.01	1.4
	Mean	3.4	33.3	11	0.28	3.1	5.3	16.66	1.55	1.07	2.77
	Minimum	2.8	24.4	9	0.27	2.5	5.2	13.47	1.04	0.96	1.4
	Maximum	4.2	34.8	13	0.44	3.8	5.4	18.77	2.23	1.2	4.11
	LSD (5 % Level)	1.52	14.88	3.2	0.06	1.11	0.19	4.41	0.87	0.51	2.74
	CV (%)	21.7	7.1	17	12.4	9.3	0.6	10.1	11.2	13.8	34.1

LSD least significant difference; CV coefficient of variation (range: 0.96–1.20 %) and sucrose: 2.77 % (range: 1.40–4.11 %). The B-lines during post-rainy season have recorded 66 % lower plant height, 40 % lower stalk yield, 10 % lower Brix% and 100 % lower sugar yield as compared to that of the rainy season. The poor performance of both female hybrid parents and cultivars during post-rainy season necessitates identifying new sources/alleles contributing to both biomass and sugar yield. The detailed characteristics of sweet sorghum cultivars and female hybrid parents adapted to post-rainy season are discussed in Chap. 4.

In summary the characterization of hybrid parental lines and cultivars helps to understand the available variability for sugars and biomass related traits in the available genotypes besides their adaptation to different seasons and further aids in stream lining the breeding objectives to improve the productivity traits in a sweet sorghum improvement program.

References

Kumar CG, Fatima A, SrinivasaRao P, Reddy BVS, Rathore A, Nageswar Rao R, Khalid S, Kamal A (2010) Characterization of improved sweet sorghum genotypes for biochemical parameters, sugar yield and its attributes at different phenological stages. Sugar Tech 12:322–328

Kumar S, Reddy KHP, Srinivasarao P, Sanjana Reddy P, Reddy BVS (2011) Study of gene effects for stalk sugar yield and its component traits in sweet sorghum [Sorghum bicolor (L.) Moench] using generation mean analysis. J Rangel Sci 1:133–142

McCallum CM, Comai L, Greene EA, Henikoff S (2000) Targeted screening for induced mutations. Nat Biotechnol 18:455–457

Reddy BVS, Sharma HC, Thakur RP, Ramesh S (2006) Characterization of ICRISAT-bred sorghum hybrid parents (Set I). Int Sorghum Millets Newsl 47(Special issue):p 135

Reddy PS, Reddy BVS, Srinivasarao P (2011) Genetic analysis of traits contributing to stalk sugar yield in sorghum. Cereal Res Commun 39:453–464

Rooney WL, Blumenthal J, Bean B, Mullet JE (2007) Designing sorghum as a dedicated bioenergy feedstock. Biofuels Bioprod Bioref 1:147–157

Srinivasarao P, Rao SS, Seetharama N, Umakanth AV, Sanjana Reddy P, Reddy BVS, Gowda CLL (2009). Sweet sorghum as a biofuel feedstock and strategies for its improvement. Information bulletin no. 77, international research institute for the semi-arid tropics (ICRISAT), p 80, ISBN 978-92-9066-518-2

Srinivasarao P, Reddy BVS, Blümmel M, Subbarao GV, Chandraraj K, Sanjana Reddy P, Parthasarathy Rao P (2010) Sweet sorghum as a biofuel feedstock: can there be food-feed-fuel trade-offs? ICID. http://www.wgcrop.icidonline.org/sweet%20sorghumdec09.pdf

Srinivasarao P, Sanjana Reddy P, Rathore A, Reddy BVS, Panwar S (2011) Application of GGE biplot and AMMI model to evaluate sweet sorghum hybrids for Genotype × Environment interaction and seasonal adaptation. Indian J Agric Sci 81:438–444

Wortmann CS, Liska AJ, Ferguson RB, Lyon DJ, Klein RM, Dweikat I (2010) Dryland performance of sweet sorghum and grain crops for biofuel. Agron J 102:319–326

Rainy Season Cultivars and Hybrid Parents

P. Srinivasa Rao, C. Ganesh Kumar, Belum V. S. Reddy,
A. Kamal, H. C. Sharma and R. P. Thakur

Abstract This chapter entitled "Rainy season cultivars and hybrid parents" gives a focussed description of improved sweet sorghum varieties/restorers, hybrids and female hybrid parents (as per PPVFRA). The coloured picture of the line is provided for easy identification. The genetic variability for all the metric traits is conspicuous and for some of the qualitative traits like anthocyanin coloration, glume color differences between the lines are not quite revealing. Among the biochemical traits such as sucrose, glucose and fructose the differences are significant among the cultivars.

Keywords Cultivars · Hybrids · Pedigree · Rainy · Kharif · Quantitative Traits · Brix%, stalk yield, sugar yield, grain yield · Glucose · Fructose · Sucrose

P. Srinivasa Rao (✉) · B. V. S. Reddy · H. C. Sharma · R. P. Thakur
International Crops Research Institute for the Semi-Arid Tropics (ICRISAT),
Patancheru 502324, India
e-mail: psrao72@gmail.com

C. Ganesh Kumar · A. Kamal
Chemical Biology Laboratory, CSIR-Indian Institute of Chemical Technology (CSIR-IICT),
Uppal Road, Hyderabad 500607, India

P. S. Rao and C. G. Kumar (eds.), *Characterization of Improved Sweet
Sorghum Cultivars*, SpringerBriefs in Agriculture,
DOI: 10.1007/978-81-322-0783-2_3, © The Author(s) 2013

1 ICSV 700 Salient Features

1. Pedigree: (IS 1082 × SC 108-3)-1-1-1-1-1.
2. Days to 50 % flowering: 84 days.
3. Plant height (m): 3.4
4. Plant girth (mm): 17.6
5. Biomass yield (t ha^{-1}): 45.6
6. Juice yield (t ha^{-1}): 17.4
7. Juice extraction (%): 38
8. Brix (%): 16
9. Sugar yield (t ha^{-1}): 2.1
10. Grain yield (t ha^{-1}): 3.1
11. Male fertility restoration (%): 85
12. Sucrose (%): 7.9
13. Glucose (%): 0.7
14. Fructose (%): 0.2
15. pH of juice: 5.0
16. Electrical conductivity of juice (mS m^{-1}).
    * Milli siemens per meter: 8.4
17. Tolerance to: Stem borer, Shootfly, Rust.
18. Adaptation: Rainy season.

Traits recorded as per guidelines for sorghum as approved by PPVFRA[a]

Characteristics	Characteristic value of candidate variety									Remarks measured value etc.
	1	2	3	4	5	6	7	8	9	
Seedling: anthocyanin colouration of coleoptile		✓								Purple
Leaf sheath: anthocyanin colouration	✓									Green
Leaf: midrib colour (5th fully developed leaf)	✓									Green
Plant: time of panicle emergence (50 % of the plants with complete panicle emergence)							✓			84
Plant: natural height of foliage up to base of flag leaf								✓		3.1
Flag leaf: yellow colouration of midrib	✓									Absent
Lemma: arista formation (awns)								✓		Strong
Stigma: anthocyanin coloration	✓									Absent
Stigma: yellow coloration					✓					Medium
Stigma length (mm)								✓		Long
Flower with pedicel: length of flower					✓					Medium
Anther: length					✓					Medium
Anther: colour of dry anther				✓						Orange
Glume: color				✓						Light red
Plant: total height								✓		3.4
Stem: diameter (at lower one-third height of plant) (mm)			✓							17.6
Leaf: length of blade of the third leaf from top including flag leaf (cm)						✓				71.6
Leaf: width of blade of the third leaf from top including flag leaf (cm)								✓		8.1
Panicle: length without peduncle			✓							18.6
Panicle: length of branches (middle third of panicle)				✓						6.1
Panicle: density at maturity (ear head compactness)						✓				Semi compact
Panicle: shape			✓							Symmetric
Neck of panicle: visible length above sheath (cm)	✓									3.8
Glume coverage (%)			✓							50
Threshability				✓						Partly threshable
Caryopsis: colour after threshing		✓								White
Grain: weight of 1000 grains (g)				✓						28.0
Grain: shape in dorsal view			✓							Circular
Grain: shape in profile view			✓							Circular
Grain: size of mark of germ						✓				Large
Grain: texture of endosperm (in longitudinal section)						✓				25 % corneous
Grain: colour of vitreous albumen			✓							Yellow
Grain: lustre						✓				Lustrous

[a] *PPVFRA* Protection of Plant Varieties and Farmers Rights Act

2 ICSV 25272 Salient Features

1. Pedigree: (DSV 4 × SSV 84)-2-2-1-2
2. Days to 50 % flowering: 74 days.
3. Plant height (cm): 3.1
4. Plant girth (mm): 14.5
5. Biomass yield (t ha^{-1}): 55.6
6. Juice yield (t ha^{-1}): 26.3
7. Juice extraction (%): 47
8. Brix (%) 17
9. Sugar yield (t ha^{-1}): 3.2
10. Grain yield (t ha^{-1}): 3.4
11. Male fertility restoration (%): 80
12. Sucrose (%): 6.4
13. Glucose (%): 1.8
14. Fructose (%): 1.1
15. pH of Juice: 5.0
16. Electrical conductivity of Juice (mS m^{-1}) * Milli siemens per meter: 9.5
17. Tolerance to: Aphids.
18. Adaptation: Rainy season.

Traits recorded as per guidelines for sorghum as approved by PPVFRA[a]

Characteristics	Characteristic value of candidate variety									Remarks measured value etc.
	1	2	3	4	5	6	7	8	9	
Seedling: anthocyanin colouration of coleoptile		√								Purple
Leaf sheath: anthocyanin colouration	√									Green
Leaf: midrib colour (5th fully developed leaf)			√							Brown
Plant: time of panicle emergence (50 % of the plants with complete panicle emergence)					√					74
Plant: natural height of foliage up to base of flag leaf							√			3
Flag leaf: yellow colouration of midrib	√									Absent
Lemma: arista formation (awns)	√									Absent
Stigma: anthocyanin coloration	√									Absent
Stigma: yellow coloration	√									Absent
Stigma length (mm)									√	Long
Flower with pedicel: length of flower					√					Medium
Anther: length							√			Long
Anther: colour of dry anther				√						Orange
Glume: color				√						Light red
Plant: total height									√	3.3
Stem: diameter (at lower one-third height of plant) (mm)		√								14.5
Leaf: length of blade of the third leaf from top including flag leaf (cm)							√			75.1
Leaf: width of blade of the third leaf from top including flag leaf (cm)							√			6.7
Panicle: length without peduncle					√					22.9
Panicle: length of branches (middle third of panicle)					√					7.8
Panicle: density at maturity (ear head compactness)					√					Semi Loose
Panicle: shape			√							Symmetric
Neck of panicle: visible length above sheath (cm)					√					14.7
Glume coverage (%)				√						50
Threshability					√					Partly threshable
Caryopsis: colour after threshing	√									White
Grain: weight of 1000 grains (g)					√					33
Grain: shape in dorsal view			√							Circular
Grain: shape in profile view			√							Circular
Grain: size of mark of germ							√			Large
Grain: texture of endosperm (in longitudinal section)							√			25 % corneous
Grain: colour of vitreous albumen			√							Yellow
Grain: lustre					√					Medium

[a] *PPVFRA* Protection of Plant Varieties and Farmers Rights Act

3 ICSV 25274 Salient Features

1. Pedigree: (DSV 4 × SSV 84)-2-5-1-3
2. Days to 50 % flowering: 87 days
3. Plant height (m): 3.5
4. Plant girth (mm): 20.45
5. Biomass yield (t ha^{-1}): 49.1
6. Juice yield (t ha^{-1}): 19.6
7. Juice extraction (%): 40
8. Brix (%): 20
9. Sugar yield (t ha^{-1}): 2.9
10. Grain yield (t ha^{-1}): 2.6
11. Male fertility restoration (%): 85
12. Sucrose (%): 9.6
13. Glucose (%): 2.0
14. Fructose (%): 0.4
15. pH of Juice: 5.0
16. Electrical conductivity of Juice (mS m^{-1}) * milli siemens per meter: 8.1
17. Tolerance to: Aphids, moderately resistant to Shoot fly and Grain mold.
18. Adaptation: Rainy season

Traits recorded as per guidelines for sorghum as approved by PPVFRA[a]

Characteristics	Characteristic value of candidate variety									Remarks measured value etc.
	1	2	3	4	5	6	7	8	9	
Seedling: anthocyanin colouration of coleoptiles	√									Purple
Leaf sheath: anthocyanin colouration		√								Green
Leaf: midrib colour (5th fully developed leaf)			√							Brown
Plant: time of panicle emergence (50 % of the plants with complete panicle emergence)		√								87
Plant: natural height of foliage up to base of flag leaf						√				3.0
Flag leaf: yellow colouration of midrib	√									Absent
Lemma: arista formation (awns)	√									Absent
Stigma: anthocyanin coloration	√									Absent
Stigma: yellow coloration	√									Absent
Stigma length (mm)					√					Medium
Flower with pedicel: length of flower					√					Medium
Anther: length					√					Medium
Anther: colour of dry anther				√						Orange
Glume: color						√				Purple
Plant: total height								√		3.5
Stem: diameter (at lower one-third height of plant) (mm)			√							19.7
Leaf: length of blade of the third leaf from top including flag leaf (cm)								√		88.7
Leaf: width of blade of the third leaf from top including flag leaf (cm)								√		9.8
Panicle: length without peduncle					√					21.3
Panicle: length of branches (middle third of panicle)					√					8.1
Panicle: density at maturity (ear head compactness)						√				Semi compact
Panicle: shape			√							Symmetric
Neck of panicle: visible length above sheath (cm)			√							8.3
Glume coverage (%)	√									25
Threshability	√									Freely threshable
Caryopsis: colour after threshing		√								White
Grain: weight of 1000 grains (g)								√		49.0
Grain: shape in dorsal view			√							Circular
Grain: shape in profile view			√							Circular
Grain: size of mark of germ						√				Large
Grain: texture of endosperm (in longitudinal section)			√							75 % corneous
Grain: colour of vitreous albumen			√							Yellow
Grain: luster						√				Lustrous

[a] *PPVFRA* Protection of Plant Varieties and Farmers Rights Act

4 ICSV 25275 Salient Features

1. Pedigree: (IS 19587 × B 24)-2-1-1-1
2. Days to 50 % flowering: 81 days
3. Plant height (m): 3.3
4. Plant girth (mm): 15.2
5. Biomass yield (t ha^{-1}): 46.6
6. Juice yield (t ha^{-1}): 11.8
7. Juice extraction (%): 23.5
8. Brix (%): 16
9. Sugar yield (t ha^{-1}): 2.8
10. Grain yield (t ha^{-1}): 2.7
11. Male fertility restoration (%): 85
12. Sucrose (%): 7.7
13. Glucose (%): 1.7
14. Fructose (%): 0.4
15. pH of Juice: 5.0
16. Electrical conductivity of juice (mS m^{-1}) * milli siemens per meter: 7.7
17. Tolerance to: Shootfly, Aphids, Rust.
18. Adaptation: Rainy season

Traits recorded as per guidelines for sorghum as approved by PPVFRA[a]

Characteristics	Characteristic value of candidate variety									Remarks measured value etc.
	1	2	3	4	5	6	7	8	9	
Seedling: anthocyanin colouration of coleoptile	√									Purple
Leaf sheath: anthocyanin colouration		√								Purple
Leaf: midrib colour (5th fully developed leaf)			√							Brown
Plant: time of panicle emergence (50 % of the plants with complete panicle emergence)							√			81 days
Plant: natural height of foliage up to base of flag leaf							√			2.8
Flag leaf: yellow colouration of midrib	√									Absent
Lemma: arista formation (awns)								√		Strong
Stigma: anthocyanin coloration	√									Absent
Stigma: yellow coloration	√									Absent
Stigma length (mm)					√					Medium
Flower with pedicel: length of flower					√					Medium
Anther: length					√					Medium
Anther: colour of dry anther				√						Dark orange
Glume: color					√					Red
Plant: total height										
Stem: diameter (at lower one-third height of plant) (mm)				√						15.2
Leaf: length of blade of the third leaf from top including flag leaf (cm)							√			72.7
Leaf: width of blade of the third leaf from top including flag leaf (cm)							√			7.9
Panicle: length without peduncle			√							17.1
Panicle: length of branches (middle third of panicle)					√					5.4
Panicle: density at maturity (ear head compactness)							√			Semi compact
Panicle: shape			√							Symmetric
Neck of panicle: visible length above sheath (cm)			√							6.7
Glume coverage (%)			√							50
Threshability		√								Partly threshable
Caryopsis: colour after threshing		√								White
Grain: weight of 1000 grains (g)					√					30.0
Grain: shape in dorsal view			√							Circular
Grain: shape in profile view			√							Circular
Grain: size of mark of germ							√			Large
Grain: texture of endosperm (in longitudinal section)							√			25 % corneous
Grain: colour of vitreous albumen			√							Yellow
Grain: lustre					√					Medium

[a] *PPVFRA* Protection of Plant Varieties and Farmers Rights Act

5 ICSV 25280 Salient Features

1. Pedigree: (ICSV 93046 × SSV 84)-7-2-1-3
2. Days to 50 % flowering: 83 days
3. Plant height (m): 3.3
4. Plant girth (mm): 15.1
5. Biomass yield (t ha^{-1}): 45.4
6. Juice yield (t ha^{-1}): 24.4
7. Juice extraction (%): 54
8. Brix (%): 19
9. Sugar yield (t ha^{-1}): 3.1
10. Grain yield (t ha^{-1}): 2.4
11. Male fertility restoration (%): 90
12. Sucrose (%): 8.9
13. Glucose (%): 1.9
14. Fructose (%): 0.4
15. pH of juice: 5.0
16. Electrical conductivity of juice (mS m^{-1}) * milli siemens per meter: 7.6
17. Tolerance to: Aphids, Shootfly, Stem borer.
18. Adaptation: Rainy season.

Traits recorded as per guidelines for sorghum as approved by PPVFRA[a]

Characteristics	Characteristic value of candidate variety									Remarks measured value etc.
	1	2	3	4	5	6	7	8	9	
Seedling: anthocyanin colouration of coleoptile	√									Purple
Leaf sheath: anthocyanin colouration		√								Purple
Leaf: midrib colour (5th fully developed leaf)	√									Green
Plant: time of panicle emergence (50 % of the plants with complete panicle emergence)							√			83 days
Plant: natural height of foliage up to base of flag leaf							√			2.7
Flag leaf: yellow colouration of midrib	√									Absent
Lemma: arista formation (awns)								√		Strong
Stigma: anthocyanin coloration	√									Absent
Stigma: yellow coloration					√					Medium
Stigma length (mm)					√					Medium
Flower with pedicel: length of flower					√					Medium
Anther: length					√					Medium
Anther: colour of dry anther				√						Dark orange
Glume: color							√			Purple
Plant: total height							√			3.3
Stem: diameter (at lower one-third height of plant) (mm)			√							15.1
Leaf: length of blade of the third leaf from top including flag leaf (cm)							√			72.4
Leaf: width of blade of the third leaf from top including flag leaf (cm)								√		8.5
Panicle: length without peduncle			√							16.8
Panicle: length of branches (middle third of panicle)					√					6.1
Panicle: density at maturity (ear head compactness)							√			Semi compact
Panicle: shape			√							Symmetric
Neck of panicle: visible length above sheath (cm)			√							6.5
Glume coverage (%)			√							50
Threshability					√					Partly threshable
Caryopsis: colour after threshing	√									White
Grain: weight of 1000 grains (g)					√					34.0
Grain: shape in dorsal view			√							Circular
Grain: shape in profile view			√							Circular
Grain: size of mark of germ							√			Large
Grain: texture of endosperm (in longitudinal section)			√							75 % corneous
Grain: colour of vitreous albumen			√							Yellow
Grain: lustre						√				Lustrous

[a] *PPVFRA* Protection of Plant Varieties and Farmers Rights Act

6 ICSV 93046 Salient Features

1. Pedigree: [((IS 1082 × SC 108 -3)-1-1-1-1-1) × (((IS 5622 × CS 3541)-20-1-1-1-1-1-1 × (UChV2 × Bulk Y-55)-1-5-1)]-9-3-1-1-1
2. Days to 50 % flowering: 86 days
3. Plant height (m): 3.6
4. Plant girth (mm): 17.9
5. Biomass yield (t ha^{-1}): 54.0
6. Juice yield (t ha^{-1}): 25.6
7. Juice extraction (%): 47
8. Brix (%): 19
9. Sugar yield (t ha^{-1}): 3.1
10. Grain yield (t ha^{-1}): 2.7
11. Male fertility restoration (%): 90
12. Sucrose (%): 9.0
13. Glucose (%): 1.6
14. Fructose (%): 0.2
15. pH of Juice: 5.0
16. Electrical conductivity of juice (mS m^{-1}) * Milli siemens per meter: 8.9
17. Tolerance to: Shootfly, Anthracnose, Grain mold and Downy mildew.
18. Adaptation: Rainy season/Post rainy season.

Traits recorded as per guidelines for sorghum as approved by PPVFRA[a]

Characteristics	Characteristic value of candidate variety									Remarks Measured value etc.
	1	2	3	4	5	6	7	8	9	
Seedling: anthocyanin colouration of coleoptile	√									Purple
Leaf sheath: anthocyanin colouration	√									Green
Leaf: midrib colour (5th fully developed leaf)		√								Brown
Plant: time of panicle emergence (50 % of the plants with complete panicle emergence)								√		86
Plant: natural height of foliage up to base of flag leaf								√		3.3
Flag leaf: yellow colouration of midrib	√									Absent
Lemma: arista formation (awns)				√						Medium
Stigma: anthocyanin coloration	√									Absent
Stigma: yellow coloration	√									Absent
Stigma length (mm)					√					Medium
Flower with pedicel: length of flower					√					Medium
Anther: length					√					Medium
Anther: colour of dry anther				√						Orange
Glume: color					√					Red
Plant: total height								√		3.6
Stem: diameter (at lower one-third height of plant) (mm)			√							17.9
Leaf: length of blade of the third leaf from top including flag leaf (cm)						√				72.8
Leaf: width of blade of the third leaf from top including flag leaf (cm)								√		8.4
Panicle: length without peduncle			√							18.1
Panicle: length of branches (middle third of panicle)				√						7.3
Panicle: density at maturity (ear head compactness)						√				Semi compact
Panicle: shape			√							Symmetric
Neck of panicle: visible length above sheath (cm)			√							5.3
Glume coverage (%)			√							50
Threshability	√									Freely threshable
Caryopsis: colour after threshing	√									White
Grain: weight of 1000 grains (g)				√						27.0
Grain: shape in dorsal view			√							Circular
Grain: shape in profile view			√							Circular
Grain: size of mark of germ							√			Large
Grain: texture of endosperm (in longitudinal section)							√			25 % corneous
Grain: colour of vitreous albumen			√							Yellow
Grain: luster							√			Lustrous

[a] *PPVFRA* Protection of Plant Varieties and Farmers Rights Act

7 SPV 422 Salient Features

1. Pedigree: Good grain 1485 (ICSV 574/ICSR 165)
2. Days to 50 % flowering: 86 days.
3. Plant height (m): 3.1
4. Plant girth (mm): 21.6
5. Biomass yield (t ha^{-1}): 47.6
6. Juice yield (t ha^{-1}): 18.7
7. Juice extraction (%): 39
8. Brix (%): 20
9. Sugar yield (t ha^{-1}): 2.8
10. Grain yield (t ha^{-1}): 3.1
11. Male fertility restoration (%): 85
12. Sucrose (%): 9.6
13. Glucose (%): 2.6
14. Fructose (%): 0.2
15. pH of juice: 5.0
16. Electrical conductivity of juice (mS m^{-1})* milli siemens per meter: 8.8
17. Tolerance to: Anthracnose, Leaf blight.
18. Adaptation: Rainy season

Traits recorded as per guidelines for sorghum as approved by PPVFRA[a]

Characteristics	Characteristic value of candidate variety									Remarks Measured value etc.
	1	2	3	4	5	6	7	8	9	
Seedling: anthocyanin colouration of coleoptile		✓								Purple
Leaf sheath: anthocyanin colouration	✓									Green
Leaf: midrib colour (5th fully developed leaf)	✓									Green
Plant: time of panicle emergence (50 % of the plants with complete panicle emergence)							✓			86
Plant: natural height of foliage up to base of flag leaf							✓			2.6
Flag leaf: yellow colouration of midrib	✓									Absent
Lemma: arista formation (awns)	✓									Absent
Stigma: anthocyanin coloration	✓									Absent
Stigma: yellow coloration	✓									Absent
Stigma length (mm)			✓							Short
Flower with pedicel: length of flower					✓					Medium
Anther: length					✓					Medium
Anther: colour of dry anther				✓						Orange
Glume: color							✓			Purple
Plant: total height								✓		3.1
Stem: diameter (at lower one-third height of plant) (mm)					✓					21.6
Leaf: length of blade of the third leaf from top including flag leaf (cm)								✓		85.2
Leaf: width of blade of the third leaf from top including flag leaf (cm)							✓			7.4
Panicle: length without peduncle			✓							16.8
Panicle: length of branches (middle third of panicle)				✓						6.2
Panicle: density at maturity (ear head compactness)							✓			Semi compact
Panicle: shape			✓							Symmetric
Neck of panicle: visible length above sheath (cm)	✓									4.3
Glume coverage (%)	✓									25
Threshability				✓						Partly threshable
Caryopsis: colour after threshing	✓									White
Grain: weight of 1000 grains (g)							✓			35.0
Grain: shape in dorsal view			✓							Circular
Grain: shape in profile view			✓							Circular
Grain: size of mark of germ							✓			Large
Grain: texture of endosperm (in longitudinal section)			✓							75 % corneous
Grain: colour of vitreous albumen			✓							Yellow
Grain: lustre							✓			Lustrous

[a] *PPVFRA* Protection of Plant Varieties and Farmers Rights Act

8 SSV 74 Salient Features

1. Pedigree: Selection from PAB 74 (Sudan; Bred by UAS, Dharwad)
2. Days to 50 % flowering: 84 days
3. Plant height (m): 3.3
4. Plant girth (mm): 20.66
5. Biomass yield (t ha^{-1}): 41.5
6. Juice yield (t ha^{-1}): 14.9
7. Juice extraction (%): 36
8. Brix (%): 16
9. Sugar yield (t ha^{-1}): 1.8
10. Grain yield (t ha^{-1}): 1.5
11. Male fertility restoration (%): 90
12. Sucrose (%): 7.0
13. Glucose (%): 1.5
14. Fructose (%): 0.3
15. pH of juice: 5.0
16. Electrical conductivity of juice (mS m^{-1}) * milli siemens per meter: 7.8
17. Tolerance to: Aphids, Rust.
18. Adaptation: Rainy season

Traits recorded as per guidelines for sorghum as approved by PPVFRA[a]

Characteristics	Characteristic value of candidate variety									Remarks measured value etc.
	1	2	3	4	5	6	7	8	9	
Seedling: anthocyanin colouration of coleoptile	✓									Grayed purple
Leaf sheath: anthocyanin colouration	✓									Grayed purple
Leaf: midrib colour (5th fully developed leaf)			✓							Yellow green
Plant: time of panicle emergence (50 % of the plants with complete panicle emergence)							✓			84
Plant: natural height of foliage up to base of flag leaf									✓	3.0
Flag leaf: yellow colouration of midrib							✓			Absent
Lemma: arista formation (awns)	✓									Absent
Stigma: anthocyanin coloration	✓									Present
Stigma: yellow coloration	✓									Absent
Stigma length (mm)			✓							Short
Flower with pedicel: length of flower					✓					Long
Anther: length					✓					Short
Anther: colour of dry anther				✓						Orange
Glume: color					✓					Grayed purple
Plant: total height									✓	3.3
Stem: diameter (at lower one-third height of plant) (mm)			✓							18.2
Leaf: length of blade of the third leaf from top including flag leaf (cm)							✓			Medium
Leaf: width of blade of the third leaf from top including flag leaf (cm)									✓	Very broad
Panicle: length without peduncle					✓					Medium
Panicle: length of branches (middle third of panicle)					✓					Medium
Panicle: density at maturity (ear head compactness)					✓					Semi loose
Panicle: shape					✓					Panicle Broader at Upper part
Neck of panicle: visible length above sheath (cm)			✓							Very short
Glume coverage (%)			✓							Short
Threshability					✓					Partly threshable
Caryopsis: colour after threshing	✓									White
Grain: weight of 1000 grains (g)							✓			36.0
Grain: shape in dorsal view			✓							Elliptic
Grain: shape in profile view			✓							Elliptic
Grain: size of mark of germ							✓			Large
Grain: texture of endosperm (in longitudinal section)							✓			25 % corneous
Grain: colour of vitreous albumen			✓							Grayed yellow
Grain: lustre							✓			Non lustrous

[a] *PPVFRA* Protection of Plant Varieties and Farmers Rights Act

9 SSV 84 Salient Features

1. Pedigree: Selection from IS 23568 (Bred by MPKV, Rahuri)
2. Days to 50 % flowering: 87 days
3. Plant height (m): 3.2
4. Plant girth (mm): 18.3
5. Biomass yield (t ha^{-1}): 38.7
6. Juice yield (t ha^{-1}): 13.5
7. Juice extraction (%): 35
8. Brix (%): 19
9. Sugar yield (t ha^{-1}): 1.9
10. Grain yield (t ha^{-1}): 3.3
11. Male fertility restoration (%): 90
12. Sucrose (%): 8.8
13. Glucose (%): 1.7
14. Fructose (%): 0.4
15. pH of juice: 5.0
16. Electrical conductivity of juice (mS m^{-1}) * milli siemens per meter:: 7.1
17. Tolerance to: Shootfly, Aphids, Rust.
18. Adaptation: Rainy season

Traits recorded as per guidelines for sorghum as approved by PPVFRA[a]

Characteristics	Characteristic value of candidate variety									Remarks Measured value etc.
	1	2	3	4	5	6	7	8	9	
Seedling: anthocyanin colouration of coleoptile	√									Yellow green
Leaf sheath: anthocyanin colouration	√									Yellow green
Leaf: midrib colour (5th fully developed leaf)			√							Yellow green
Plant: time of panicle emergence (50 % of the plants with complete panicle emergence)								√		87
Plant: natural height of foliage up to base of flag leaf								√		2.9
Flag leaf: yellow colouration of midrib	√									Absent
Lemma: arista formation (awns)	√									Absent
Stigma: anthocyanin coloration	√									Present
Stigma: yellow coloration	√									Present
Stigma length (mm)			√							Medium
Flower with pedicel: length of flower					√					Long
Anther: length					√					Short
Anther: colour of dry anther				√						Grayed orange
Glume: color							√			Grayed purple
Plant: total height								√		3.2
Stem: diameter (at lower one-third height of plant) (mm)			√							18.3
Leaf: length of blade of the third leaf from top including flag leaf (cm)								√		Long
Leaf: width of blade of the third leaf from top including flag leaf (cm)								√		Very broad
Panicle: length without peduncle			√							Short
Panicle: length of branches (middle third of panicle)					√					Medium
Panicle: density at maturity (ear head compactness)							√			Semi compact
Panicle: shape			√							Panicle Braoder at Upper part
Neck of panicle: visible length above sheath (cm)	√									Absent
Glume coverage (%)			√							Short
Threshability					√					Partly threshable
Caryopsis: colour after threshing	√									Yellow White
Grain: weight of 1000 grains (g)				√						25.0
Grain: shape in dorsal view			√							Circular
Grain: shape in profile view			√							Circular
Grain: size of mark of germ							√			Large
Grain: texture of endosperm (in longitudinal section)							√			25 % corneous
Grain: colour of vitreous albumen			√							Grayed yellow
Grain: lustre							√			Non lustrous

[a] *PPVFRA* Protection of Plant Varieties and Farmers Rights Act

10 RSSV 9 (CSV 19SS) Salient Features

1. Pedigree: (RSSV 2 X SPV 462)-2-1-1 (Bred by MPKV, Rahuri)
2. Days to 50 % flowering: 85 days
3. Plant height (m): 3.3
4. Plant girth (mm): 17.8
5. Biomass yield (t ha^{-1}): 58.9
6. Juice yield (t ha^{-1}): 21.4
7. Juice extraction (%): 36
8. Brix (%): 16
9. Sugar yield (t ha^{-1}): 2.6
10. Grain yield (t ha^{-1}): 1.5
11. Male fertility restoration (%): 90
12. Sucrose (%): 9.0
13. Glucose (%): 1.7
14. Fructose (%): 2.8
15. pH of Juice: 4.0
16. Electrical conductivity of juice (mS m^{-1}) * milli siemens per meter: 9.4
17. Tolerance to: Aphids, Shootfly.
18. Adaptation: Rainy season

Traits recorded as per guidelines for sorghum as approved by PPVFRA[a]

Characteristics	Characteristic value of candidate variety									Remarks measured value etc.
	1	2	3	4	5	6	7	8	9	
Seedling: anthocyanin colouration of coleoptile		√								Grayed purple
Leaf sheath: anthocyanin colouration	√									Grayed purple
Leaf: midrib colour (5th fully developed leaf)	√									Yellow green
Plant: time of panicle emergence (50 % of the plants with complete panicle emergence)							√			85
Plant: natural height of foliage up to base of flag leaf								√		2.9
Flag leaf: yellow colouration of midrib	√									Absent
Lemma: arista formation (awns)								√		Present
Stigma: anthocyanin coloration	√									Present
Stigma: yellow coloration					√					Present
Stigma length (mm)								√		Long
Flower with pedicel: length of flower					√					Long
Anther: length					√					Medium
Anther: colour of dry anther				√						Grayed orange
Glume: color						√				Grayed purple
Plant: total height								√		3.3
Stem: diameter (at lower one-third height of plant) (mm)			√							17.8
Leaf: length of blade of the third leaf from top including flag leaf (cm)						√				Medium
Leaf: width of blade of the third leaf from top including flag leaf (cm)								√		Broad
Panicle: length without peduncle			√							Short
Panicle: length of branches (middle third of panicle)				√						Medium
Panicle: density at maturity (ear head compactness)						√				Semi compact
Panicle: shape			√							Panicle broader in lower part
Neck of panicle: visible length above sheath (cm)	√									Absent or Very short
Glume coverage (%)			√							Medium
Threshability					√					Difficult threshable
Caryopsis: colour after threshing				√						Yellow white
Grain: weight of 1000 grains (g)					√					26.0
Grain: shape in dorsal view			√							Circular
Grain: shape in profile view			√							Circular
Grain: size of mark of germ					√					Large
Grain: texture of endosperm (in longitudinal section)					√					50 % corneous
Grain: colour of vitreous albumen			√							Grayed yellow
Grain: lustre				√						Non lustrous

[a] *PPVFRA* Protection of Plant Varieties and Farmers Rights Act

11 ICSB 38 Salient Features

1. Pedigree: [(BTx 623 × MR 862)B lines bulk]-5-1-3-5
2. Days to 50 % flowering: 74 days
3. Plant height (m): 1.4
4. Plant girth (mm): 17.7
5. Biomass yield (t ha^{-1}): 16.0
6. Juice yield (t ha^{-1}): 4.4
7. Juice extraction (%): 27.5
8. Brix (%): 9
9. Sugar yield (t ha^{-1}): 0.3
10. Grain yield (t ha^{-1}): 2.3
11. Male fertility restoration (%): 80
12. Sucrose (%): 4.7
13. Glucose (%): 2.1
14. Fructose (%): 0.7
15. pH of juice: 5.5
16. Electrical conductivity of juice (mS m^{-1}) * Milli siemens per meter: 14.7
17. Tolerance to: Shoot fly, Stem borer, Rust.
18. Adaptation: Rainy season.

Traits recorded as per guidelines for sorghum as approved by PPVFRA[a]

Characteristics	Characteristic value of candidate hybrid parent									Remarks measured value etc.
	1	2	3	4	5	6	7	8	9	
Seedling: anthocyanin colouration of coleoptile	√									Green
Leaf sheath: anthocyanin colouration	√									Green
Leaf: midrib colour (5th fully developed leaf)			√							Brown
Plant: time of panicle emergence (50 % of the plants with complete panicle emergence)						√				74
Plant: natural height of foliage up to base of flag leaf		√								1.1
Flag leaf: yellow colouration of midrib	√									Absent
Lemma: arista formation (awns)	√									Absent
Stigma: anthocyanin coloration	√									Absent
Stigma: yellow coloration	√									Average
Stigma length (mm)			√							Short
Flower with pedicel: length of flower			√							Short
Anther: length			√							Short
Anther: colour of dry anther				√						Orange
Glume: color				√						Light red
Plant: total height			√							1.4
Stem: diameter (at lower one-third height of plant) (mm)						√				17.7
Leaf: length of blade of the third leaf from top including flag leaf (cm)								√		80.6
Leaf: width of blade of the third leaf from top including flag leaf (cm)						√				7.6
Panicle: length without peduncle					√					30.7
Panicle: length of branches (middle third of panicle)					√					7.7
Panicle: density at maturity (ear head compactness)						√				Semi compact
Panicle: shape			√							Symmetric
Neck of panicle: visible length above sheath (cm)							√			24
Glume coverage (%)	√									25
Threshability					√					Partly threshable
Caryopsis: colour after threshing	√									White
Grain: weight of 1000 grains (g)					√					26.0
Grain: shape in dorsal view			√							Circular
Grain: shape in profile view			√							Circular
Grain: size of mark of germ							√			Large
Grain: texture of endosperm (in longitudinal section)				√						75 % corneous
Grain: colour of vitreous albumen				√						Yellow
Grain: lustre	√									Non-lustrous

[a] *PPVFRA* Protection of Plant Varieties and Farmers Rights Act

12 ICSB 474 Salient Features

1. Pedigree: (IS 18432 × ICSB 6)-11-1-1-2-2
2. Days to 50 % flowering: 70 days
3. Plant height (m): 2.1
4. Plant girth (mm): 14.1
5. Biomass yield (t ha^{-1}): 23.1
6. Juice yield (t ha^{-1}): 8.0
7. Juice extraction (%): 34.4
8. Brix (%): 11
9. Sugar yield (t ha^{-1}): 0.6
10. Grain yield (t ha^{-1}): 1.4
11. Male fertility restoration (%): 75
12. Sucrose (%): 5.7
13. Glucose (%): 1.2
14. Fructose (%): 0.7
15. pH of juice: 5.1
16. Electrical conductivity of juice (mS m^{-1}) * milli siemens per meter: 10.2
17. Tolerance to: Stem borer
18. Adaptation: Rainy season

Traits recorded as per guidelines for sorghum as approved by PPVFRA[a]

Characteristics	Characteristic value of candidate hybrid parent									Remarks Measured value etc.
	1	2	3	4	5	6	7	8	9	
Seedling: anthocyanin colouration of coleoptile		√								Purple
Leaf sheath: anthocyanin colouration	√									Green
Leaf: midrib colour (5th fully developed leaf)		√								Green
Plant: time of panicle emergence (50 % of the plants with complete panicle emergence)							√			70
Plant: natural height of foliage up to base of flag leaf					√					1.8
Flag leaf: yellow colouration of midrib	√									Absent
Lemma: arista formation (awns)								√		Strong
Stigma: anthocyanin coloration	√									Absent
Stigma: yellow coloration				√						Medium
Stigma length (mm)				√						Medium
Flower with pedicel: length of flower				√						Medium
Anther: length				√						Medium
Anther: colour of dry anther				√						Orange
Glume: color				√						Red
Plant: total height						√				2.1
Stem: diameter (at lower one-third height of plant) (mm)						√				14.1
Leaf: length of blade of the third leaf from top including flag leaf (cm)						√				73.1
Leaf: width of blade of the third leaf from top including flag leaf (cm)						√				7.3
Panicle: length without peduncle					√					21
Panicle: length of branches (middle third of panicle)					√					6
Panicle: density at maturity (ear head compactness)						√				Semi compact
Panicle: shape			√							Symmetric
Neck of panicle: visible length above sheath (cm)					√					11.4
Glume coverage (%)			√							50
Threshability	√									Partly threshable
Caryopsis: colour after threshing	√									White
Grain: weight of 1000 grains (g)						√				43.0
Grain: shape in dorsal view			√							Circular
Grain: shape in profile view			√							Circular
Grain: size of mark of germ						√				Large
Grain: texture of endosperm (in longitudinal section)			√							75 % corneous
Grain: colour of vitreous albumen			√							Yellow
Grain: lustre					√					Medium

[a] *PPVFRA* Protection of Plant Varieties and Farmers Rights Act

13 ICSB 675 Salient Features

1. Pedigree: (E 36-1 × ICSB 17) -12-2
2. Days to 50 % flowering: 76 days.
3. Plant height (m): 1.2
4. Plant girth (mm): 15.0
5. Biomass yield (t ha^{-1}): 21.4
6. Juice yield (t ha^{-1}): 6.7
7. Juice extraction (%): 31.3
8. Brix (%): 12
9. Sugar yield (t ha^{-1}): 0.6
10. Grain yield (t ha^{-1}): 0.6
11. Male fertility restoration (%): 60
12. Sucrose (%): 1.7
13. Glucose (%): 2.4
14. Fructose (%): 0.7
15. pH of juice: 5.3
16. Electrical conductivity of juice (mS m^{-1}) * Milli siemens per meter: 12.2
17. Tolerance to: Aphids, terminal drought.
18. Adaptation: Rainy season

Traits recorded as per guidelines for sorghum as approved by PPVFRA[a]

Characteristics	Characteristic value of candidate hybrid parent									Remarks Measured value etc.
	1	2	3	4	5	6	7	8	9	
Seedling: anthocyanin colouration of coleoptile	✓									Purple
Leaf sheath: anthocyanin colouration		✓								Purple
Leaf: midrib colour (5th fully developed leaf)			✓							Brown
Plant: time of panicle emergence (50 % of the plants with complete panicle emergence)						✓				76
Plant: natural height of foliage up to base of flag leaf		✓								1.0
Flag leaf: yellow colouration of midrib	✓									Absent
Lemma: arista formation (awns)	✓									Absent
Stigma: anthocyanin coloration	✓									Absent
Stigma: yellow coloration	✓									Average
Stigma length (mm)			✓							Short
Flower with pedicel: length of flower				✓						Medium
Anther: length				✓						Medium
Anther: colour of dry anther			✓							Orange
Glume: color						✓				Purple
Plant: total height		✓								1.2
Stem: diameter (at lower one-third height of plant) (mm)						✓				15.0
Leaf: length of blade of the third leaf from top including flag leaf (cm)							✓			85.3
Leaf: width of blade of the third leaf from top including flag leaf (cm)							✓			8.5
Panicle: length without peduncle				✓						21
Panicle: length of branches (middle third of panicle)				✓						7.4
Panicle: density at maturity (ear head compactness)						✓				Semi compact
Panicle: shape			✓							Symmetric
Neck of panicle: visible length above sheath (cm)				✓						12.7
Glume coverage (%)	✓									25
Threshability	✓									Freely threshable
Caryopsis: colour after threshing	✓									White
Grain: weight of 1000 grains (g)			✓							16.0
Grain: shape in dorsal view			✓							Circular
Grain: shape in profile view			✓							Circular
Grain: size of mark of germ						✓				Large
Grain: texture of endosperm (in longitudinal section)			✓							75 % corneous
Grain: colour of vitreous albumen			✓							Yellow
Grain: lustre				✓						Medium

[a] *PPVFRA* Protection of Plant Varieties and Farmers Rights Act

14 ICSB 702 Salient Features

1. Pedigree: (ICSB 101 × SP 36257)-3-3-1-2
2. Days to 50 % flowering: 67 days.
3. Plant height (m): 1.6
4. Plant girth (mm): 15.0
5. Biomass yield (t ha^{-1}): 17.4
6. Juice yield (t ha^{-1}): 7.6
7. Juice extraction (%): 43.6
8. Brix (%): 13
9. Sugar yield (t ha^{-1}): 0.74
10. Grain yield (t ha^{-1}): 4.1
11. Male fertility restoration (%): 70
12. Sucrose (%): 4.8
13. Glucose (%): 1.0
14. Fructose (%): 0.6
15. pH of juice: 5.3
16. Electrical conductivity of juice (mS m^{-1}) * Milli siemens per meter: 11.5
17. Tolerance to: Aphids.
18. Adaptation: Rainy season.

Traits recorded as per guidelines for sorghum as approved by PPVFRA[a]

Characteristics	Characteristic value of candidate hybrid parent									Remarks Measured value etc.
	1	2	3	4	5	6	7	8	9	
Seedling: anthocyanin colouration of coleoptile	√									Green
Leaf sheath: anthocyanin colouration	√									Green
Leaf: midrib colour (5th fully developed leaf)		√								Brown
Plant: time of panicle emergence (50 % of the plants with complete panicle emergence)						√				67
Plant: natural height of foliage up to base of flag leaf					√					1.4
Flag leaf: yellow colouration of midrib	√									Absent
Lemma: arista formation (awns)	√									Absent
Stigma: anthocyanin coloration	√									Absent
Stigma: yellow coloration	√									Average
Stigma length (mm)		√								Short
Flower with pedicel: length of flower		√								Short
Anther: length			√							Short
Anther: colour of dry anther			√							Orange
Glume: color			√							Light red
Plant: total height			√							1.6
Stem: diameter (at lower one-third height of plant) (mm)					√					15.0
Leaf: length of blade of the third leaf from top including flag leaf (cm)					√					80.9
Leaf: width of blade of the third leaf from top including flag leaf (cm)							√			7.8
Panicle: length without peduncle		√								24.1
Panicle: length of branches (middle third of panicle)			√							7.4
Panicle: density at maturity (ear head compactness)					√					Semi compact
Panicle: shape										Symmetric
Neck of panicle: visible length above sheath (cm)										15.7
Glume coverage (%)										25
Threshability										Partly threshable
Caryopsis: colour after threshing			√							Light yellow
Grain: weight of 1000 grains (g)			√							37.0
Grain: shape in dorsal view		√								Circular
Grain: shape in profile view		√								Circular
Grain: size of mark of germ					√					Large
Grain: texture of endosperm (in longitudinal section)		√								75 % corneous
Grain: colour of vitreous albumen		√								Yellow
Grain: lustre			√							Medium

[a] *PPVFRA* Protection of Plant Varieties and Farmers Rights Act

15 ICSB 724 Salient Features

1. Pedigree: ICSP 1B/R MFR-S 7-303-2-1
2. Days to 50 % flowering: 76 days
3. Plant height (m): 1.9
4. Plant girth (mm): 15.0
5. Biomass yield (t ha^{-1}): 24.8
6. Juice yield (t ha^{-1}): 9.7
7. Juice extraction (%): 39.1
8. Brix (%): 13
9. Sugar yield (t ha^{-1}): 0.9
10. Grain yield (t ha^{-1}): 4.3
11. Male fertility restoration (%): 75
12. Sucrose (%): 4.6
13. Glucose (%): 1.4
14. Fructose (%): 0.4
15. pH of juice: 5.3
16. Electrical conductivity of juice (mS m^{-1}) * Milli siemens per meter: 9.7
17. Tolerance to: Aphids
18. Adaptation: Rainy season.

Traits recorded as per guidelines for sorghum as approved by PPVFRA[a]

Characteristics	Characteristic value of candidate hybrid parent									Remarks Measured value etc.
	1	2	3	4	5	6	7	8	9	
Seedling: anthocyanin colouration of coleoptile		√								Purple
Leaf sheath: anthocyanin colouration	√									Green
Leaf: midrib colour (5th fully developed leaf)			√							Green
Plant: time of panicle emergence (50 % of the plants with complete panicle emergence)							√			76
Plant: natural height of foliage up to base of flag leaf					√					1.7
Flag leaf: yellow colouration of midrib	√									Absent
Lemma: arista formation (awns)	√									Absent
Stigma: anthocyanin coloration	√									Absent
Stigma: yellow coloration	√									Average
Stigma length (mm)			√							Short
Flower with pedicel: length of flower			√							Short
Anther: length					√					Medium
Anther: colour of dry anther			√							Orange
Glume: color			√							Light red
Plant: total height						√				1.9
Stem: diameter (at lower one-third height of plant) (mm)							√			15.0
Leaf: length of blade of the third leaf from top including flag leaf (cm)								√		78.4
Leaf: width of blade of the third leaf from top including flag leaf (cm)								√		9.1
Panicle: length without peduncle					√					20.7
Panicle: length of branches (middle third of panicle)					√					6.5
Panicle: density at maturity (ear head compactness)							√			Semi compact
Panicle: shape			√							Symmetric
Neck of panicle: visible length above sheath (cm)					√					8.2
Glume coverage (%)	√									25
Threshability					√					Partly threshable
Caryopsis: colour after threshing			√							Light yellow
Grain: weight of 1000 grains (g)					√					27.0
Grain: shape in dorsal view			√							Circular
Grain: shape in profile view			√							Circular
Grain: size of mark of germ					√					Medium
Grain: texture of endosperm (in longitudinal section)							√			25 % corneous
Grain: colour of vitreous albumen			√							Yellow
Grain: lustre							√			Lustrous

[a] *PPVFRA* Protection of Plant Varieties and Farmers Rights Act

16 ICSB 731 Salient Features

1. Pedigree: ICSV 117 1BF
2. Days to 50 % flowering: 74 days
3. Plant height (m): 2.0
4. Plant girth (mm): 16.4
5. Biomass yield (t ha^{-1}): 20.3
6. Juice yield (t ha^{-1}): 9.2
7. Juice extraction (%): 45
8. Brix (%): 14
9. Sugar yield (t ha^{-1}): 1.0
10. Grain yield (t ha^{-1}): 3.9
11. Male fertility restoration (%): 70
12. Sucrose (%): 6.9
13. Glucose (%): 2.9
14. Fructose (%): 1.3
15. pH of juice: 5.1
16. Electrical conductivity of Juice (mS m^{-1}) * Milli siemens per meter: 9.1
17. Tolerance to: Anthracnose.
18. Adaptation: Rainy season

Traits recorded as per guidelines for sorghum as approved by PPVFRA[a]

Characteristics	Characteristic value of candidate hybrid parent									Remarks Measured value etc.
	1	2	3	4	5	6	7	8	9	
Seedling: anthocyanin colouration of coleoptile		√								Purple
Leaf sheath: anthocyanin colouration	√									Green
Leaf: midrib colour (5th fully developed leaf)			√							Brown
Plant: time of panicle emergence (50 % of the plants with complete panicle emergence)						√				74
Plant: natural height of foliage up to base of flag leaf				√						1.8
Flag leaf: yellow colouration of midrib	√									Absent
Lemma: arista formation (awns)	√									Absent
Stigma: anthocyanin coloration	√									Absent
Stigma: yellow coloration	√									Average
Stigma length (mm)			√							Short
Flower with pedicel: length of flower			√							Short
Anther: length				√						Medium
Anther: colour of dry anther				√						Orange
Glume: color			√							Light red
Plant: total height						√				2.0
Stem: diameter (at lower one-third height of plant) (mm)			√							16.4
Leaf: length of blade of the third leaf from top including flag leaf (cm)								√		88.6
Leaf: width of blade of the third leaf from top including flag leaf (cm)								√		8.1
Panicle: length without peduncle				√						22.1
Panicle: length of branches (middle third of panicle)			√							6.1
Panicle: density at maturity (ear head compactness)							√			Semi compact
Panicle: shape			√							Symmetric
Neck of panicle: visible length above sheath (cm)				√						10.8
Glume coverage (%)			√							25
Threshability				√						Partly threshable
Caryopsis: colour after threshing	√									White
Grain: weight of 1000 grains (g)				√						29.0
Grain: shape in dorsal view			√							Circular
Grain: shape in profile view			√							Circular
Grain: size of mark of germ				√						Medium
Grain: texture of endosperm (in longitudinal section)			√							75 % corneous
Grain: colour of vitreous albumen			√							Yellow
Grain: lustre				√						Medium

[a] *PPVFRA* Protection of Plant Varieties and Farmers Rights Act

17 ICSSH 28 Salient Features

1. Pedigree: ICSA 474 × SSV 74
2. Days to 50 % flowering: 76 days
3. Plant height (m): 3.8
4. Plant girth (mm): 17.5
5. Biomass yield (t ha^{-1}): 63.6
6. Juice yield (t ha^{-1}): 28.3
7. Juice extraction (%): 45
8. Brix (%): 16
9. Sugar yield (t ha^{-1}): 3.4
10. Grain yield (t ha^{-1}): 2.4
11. Male fertility restoration (%): 82
12. Sucrose (%): 6.3
13. Glucose (%): 1.5
14. Fructose (%): 0.5
15. pH of juice: 5.0
16. Electrical conductivity of juice (mS m^{-1}) * milli siemens per meter: 9.6
17. Tolerance to: Aphids, Shootfly.
18. Adaptation: Rainy season/post rainy season.

Traits recorded as per guidelines for sorghum as approved by PPVFRA[a]

Characteristics	Characteristic value of candidate hybrid									Remarks Measured value etc.
	1	2	3	4	5	6	7	8	9	
Seedling: anthocyanin colouration of coleoptile		√								Purple
Leaf sheath: anthocyanin colouration	√									Green
Leaf: midrib colour (5th fully developed leaf)	√									White
Plant: time of panicle emergence (50 % of the plants with complete panicle emergence)							√			76
Plant: natural height of foliage up to base of flag leaf							√			3.5
Flag leaf: yellow colouration of midrib		√								Absent
Lemma: arista formation (awns)		√								Absent
Stigma: anthocyanin coloration		√								Absent
Stigma: yellow coloration					√					Medium
Stigma length (mm)			√							Short
Flower with pedicel: length of flower							√			Long
Anther: length					√					Medium
Anther: colour of dry anther				√						Orange
Glume: color					√					Red
Plant: total height							√			3.8
Stem: diameter (at lower one-third height of plant) (mm)							√			17.5
Leaf: length of blade of the third leaf from top including flag leaf (cm)								√		81.1
Leaf: width of blade of the third leaf from top including flag leaf (cm)								√		9.8
Panicle: length without peduncle					√					24.2
Panicle: length of branches (middle third of panicle)					√					7.9
Panicle: density at maturity (ear head compactness)					√					Semi loose
Panicle: shape			√							Symmetric
Neck of panicle: visible length above sheath (cm)					√					11.7
Glume coverage (%)	√									25
Threshability	√									Freely threshable
Caryopsis: colour after threshing	√									White
Grain: weight of 1000 grains (g)								√		47.0
Grain: shape in dorsal view			√							Circular
Grain: shape in profile view			√							Circular
Grain: size of mark of germ						√				Large
Grain: texture of endosperm (in longitudinal section)			√							75 % corneous
Grain: colour of vitreous albumen			√							Yellow
Grain: lustre	√									Non-Lustrous

[a] *PPVFRA* Protection of Plant Varieties and Farmers Rights Act

18 ICSSH 29 Salient Features

1. Pedigree: ICSA 675 × SSV 74
2. Days to 50 % flowering: 82 days
3. Plant height (m): 3.8
4. Plant girth (mm): 17.8
5. Biomass yield (t ha^{-1}): 58.5
6. Juice yield (t ha^{-1}): 25.4
7. Juice extraction (%): 43
8. Brix (%): 16
9. Sugar yield (t ha^{-1}): 3.6
10. Grain yield (t ha^{-1}): 2.8
11. Male fertility restoration (%): 82
12. Sucrose (%): 7.6
13. Glucose (%): 0.4
14. Fructose (%): 1.5
15. pH of juice: 5.0
16. Electrical conductivity of juice (mS m^{-1}) * milli siemens per meter: 7.5
17. Tolerance to: Anthracnose.
18. Adaptation: Rainy season

Traits recorded as per guidelines for sorghum as approved by PPVFRA[a]

Characteristics	Characteristic value of candidate hybrid									Remarks measured value etc.
	1	2	3	4	5	6	7	8	9	
Seedling: anthocyanin colouration of coleoptile		√								Purple
Leaf sheath: anthocyanin colouration			√							Purple
Leaf: midrib colour (5th fully developed leaf)		√								Green
Plant: time of panicle emergence (50 % of the plants with complete panicle emergence)						√				82
Plant: natural height of foliage up to base of flag leaf								√		3.5
Flag leaf: yellow colouration of midrib	√									Absent
Lemma: arista formation (awns)	√									Absent
Stigma: anthocyanin coloration	√									Absent
Stigma: yellow coloration	√									Absent
Stigma length (mm)			√							Short
Flower with pedicel: length of flower					√					Medium
Anther: length					√					Medium
Anther: colour of dry anther				√						Orange
Glume: color						√				Purple
Plant: total height						√				3.8
Stem: diameter (at lower one-third height of plant) (mm)						√				17.8
Leaf: length of blade of the third leaf from top including flag leaf (cm)								√		91.4
Leaf: width of blade of the third leaf from top including flag leaf (cm)								√		9.9
Panicle: length without peduncle					√					25.8
Panicle: length of branches (middle third of panicle)					√					7.4
Panicle: density at maturity (ear head compactness)							√			Semi compact
Panicle: shape			√							Symmetric
Neck of panicle: visible length above sheath (cm)			√							9.3
Glume coverage (%)	√									25
Threshability					√					Partly threshable
Caryopsis: colour after threshing				√						Light yellow
Grain: weight of 1000 grains (g)					√					27.0
Grain: shape in dorsal view			√							Circular
Grain: shape in profile view			√							Circular
Grain: size of mark of germ							√			Large
Grain: texture of endosperm (in longitudinal section)			√							75 % corneous
Grain: colour of vitreous albumen			√							Yellow
Grain: lustre	√									Non-lustrous

[a] *PPVFRA* Protection of Plant Varieties and Farmers Rights Act

19 ICSSH 30 Salient Features

1. Pedigree: ICSA 724 × SSV 74
2. Days to 50 % flowering: 77 days
3. Plant height (m): 3.5
4. Plant girth (mm): 18.8
5. Biomass yield (t ha^{-1}): 52.0
6. Juice yield (t ha^{-1}): 24.5
7. Juice extraction (%): 47
8. Brix (%): 18
9. Sugar yield (t ha^{-1}): 3.3
10. Grain yield (t ha^{-1}): 2.2
11. Male fertility restoration (%): 92
12. Sucrose (%): 7.8
13. Glucose (%): 0.6
14. Fructose (%): 2.0
15. pH of juice: 5.0
16. Electrical conductivity of juice (mS m^{-1}) * Milli siemens per meter: 7.5
17. Tolerance to: Anthracnose, Rust.
18. Adaptation: Rainy season

Traits recorded as per guidelines for sorghum as approved by PPVFRA[a]

Characteristics	1	2	3	4	5	6	7	8	9	Remarks measured value etc.
Seedling: anthocyanin colouration of coleoptile		✓								Purple
Leaf sheath: anthocyanin colouration	✓									Green
Leaf: midrib colour (5th fully developed leaf)		✓								Green
Plant: time of panicle emergence (50 % of the plants with complete panicle emergence)							✓			77
Plant: natural height of foliage up to base of flag leaf							✓			3.2
Flag leaf: yellow colouration of midrib	✓									Absent
Lemma: arista formation (awns)	✓									Absent
Stigma: anthocyanin coloration	✓									Absent
Stigma: yellow coloration		✓								Absent
Stigma length (mm)			✓							Short
Flower with pedicel: length of flower			✓							Short
Anther: length					✓					Medium
Anther: colour of dry anther				✓						Orange
Glume: color					✓					Red
Plant: total height							✓			3.2
Stem: diameter (at lower one-third height of plant) (mm)							✓			18.8
Leaf: length of blade of the third leaf from top including flag leaf (cm)								✓		83.3
Leaf: width of blade of the third leaf from top including flag leaf (cm)								✓		9.9
Panicle: length without peduncle					✓					23.8
Panicle: length of branches (middle third of panicle)					✓					8
Panicle: density at maturity (ear head compactness)					✓					Semi loose
Panicle: shape			✓							Symmetric
Neck of panicle: visible length above sheath (cm)			✓							12.9
Glume coverage (%)	✓									25
Threshability					✓					Partly threshable
Caryopsis: colour after threshing				✓						Light yellow
Grain: weight of 1000 grains (g)							✓			44.0
Grain: shape in dorsal view			✓							Circular
Grain: shape in profile view			✓							Circular
Grain: size of mark of germ							✓			Large
Grain: texture of endosperm (in longitudinal section)				✓						75 % corneous
Grain: colour of vitreous albumen				✓						Yellow
Grain: lustre					✓					Medium

[a] *PPVFRA* Protection of Plant Varieties and Farmers Rights Act

20 ICSSH 31 Salient Features

1. Pedigree: ICSA 38 × ICSV 700
2. Days to 50 % flowering: 75 days.
3. Plant height (m): 3.1
4. Plant girth (mm): 14.2
5. Biomass yield (t ha^{-1}): 51.2
6. Juice yield (t ha^{-1}): 22.4
7. Juice extraction (%): 44
8. Brix (%): 16
9. Sugar yield (t ha^{-1}): 2.7
10. Grain yield (t ha^{-1}): 3.3
11. Male fertility restoration (%): 91
12. Sucrose (%): 8.4
13. Glucose (%): 0.2
14. Fructose (%): 1.4
15. pH of juice: 5.0
16. Electrical conductivity of juice (mS m^{-1}) * milli siemens per meter: 9.3
17. Tolerance to: Aphids, Anthracnose, Shootfly, Rust
18. Adaptation: Rainy season

Traits recorded as per guidelines for sorghum as approved by PPVFRA[a]

Characteristics	Characteristic value of candidate hybrid									Remarks measured value etc.
	1	2	3	4	5	6	7	8	9	
Seedling: anthocyanin colouration of coleoptile	√									Purple
Leaf sheath: anthocyanin colouration	√									Green
Leaf: midrib colour (5th fully developed leaf)			√							Brown
Plant: time of panicle emergence (50 % of the plants with complete panicle emergence)				√						75
Plant: natural height of foliage up to base of flag leaf							√			2.9
Flag leaf: yellow colouration of midrib	√									Absent
Lemma: arista formation (awns)	√									Absent
Stigma: anthocyanin coloration	√									Absent
Stigma: yellow coloration	√									Absent
Stigma length (mm)			√							Short
Flower with pedicel: length of flower					√					Medium
Anther: length					√					Medium
Anther: colour of dry anther				√						Orange
Glume: color					√					Red
Plant: total height							√			3.1
Stem: diameter (at lower one-third height of plant) (mm)							√			14.2
Leaf: length of blade of the third leaf from top including flag leaf (cm)								√		85.3
Leaf: width of blade of the third leaf from top including flag leaf (cm)								√		8.6
Panicle: length without peduncle					√					27.5
Panicle: length of branches (middle third of panicle)					√					7.6
Panicle: density at maturity (ear head compactness)					√					Semi loose
Panicle: shape			√							Symmetric
Neck of panicle: visible length above sheath (cm)					√					14.9
Glume coverage (%)					√					75
Threshability					√					Partly threshable
Caryopsis: colour after threshing	√									White
Grain: weight of 1000 grains (g)					√					32.0
Grain: shape in dorsal view			√							Circular
Grain: shape in profile view			√							Circular
Grain: size of mark of germ							√			Large
Grain: texture of endosperm (in longitudinal section)			√							75 % corneous
Grain: colour of vitreous albumen			√							Yellow
Grain: lustre					√					Medium

[a] *PPVFRA* Protection of Plant Varieties and Farmers Rights Act

21 ICSSH 39 Salient Features

1. Pedigree: ICSA 702 × SSV 74
2. Days to 50 % flowering: 74 days.
3. Plant height (m): 3.3
4. Plant girth (mm): 16.4
5. Biomass yield (t ha^{-1}): 46.9
6. Juice yield (t ha^{-1}): 22.5
7. Juice extraction (%): 48
8. Brix (%): 18
9. Sugar yield (t ha^{-1}): 3.0
10. Grain yield (t ha^{-1}): 2.2
11. Male fertility restoration (%): 83
12. Sucrose (%): 8.1
13. Glucose (%): 1.1
14. Fructose (%): 0.4
15. pH of juice: 5.0
16. Electrical conductivity of juice (mS m^{-1}) * milli siemens per meter: 9.5
17. Tolerance to: Aphids, Anthracnose, Rust
18. Adaptation: Rainy season

Traits recorded as per guidelines for sorghum as approved by PPVFRA*

Characteristics	Characteristic value of candidate hybrid									Remarks measured value etc.
	1	2	3	4	5	6	7	8	9	
Seedling: anthocyanin colouration of coleoptile	√									Green
Leaf sheath: anthocyanin colouration	√									Green
Leaf: midrib colour (5th fully developed leaf)		√								Green
Plant: time of panicle emergence (50 % of the plants with complete panicle emergence)				√						74
Plant: natural height of foliage up to base of flag leaf						√				3.0
Flag leaf: yellow colouration of midrib	√									Absent
Lemma: arista formation (awns)	√									Absent
Stigma: anthocyanin coloration	√									Absent
Stigma: yellow coloration	√									Absent
Stigma length (mm)					√					Medium
Flower with pedicel: length of flower					√					Medium
Anther: length					√					Medium
Anther: colour of dry anther				√						Orange
Glume: color					√					Red
Plant: total height							√			3.3
Stem: diameter (at lower one-third height of plant) (mm)							√			16.4
Leaf: length of blade of the third leaf from top including flag leaf (cm)								√		86.4
Leaf: width of blade of the third leaf from top including flag leaf (cm)								√		9.5
Panicle: length without peduncle					√					25.8
Panicle: length of branches (middle third of panicle)					√					9.1
Panicle: density at maturity (ear head compactness)					√					Semi loose
Panicle: shape			√							Symmetric
Neck of panicle: visible length above sheath (cm)			√							9.3
Glume coverage (%)		√								25
Threshability		√								Freely threshable
Caryopsis: colour after threshing				√						Light yellow
Grain: weight of 1000 grains (g)					√					36.0
Grain: shape in dorsal view			√							Circular
Grain: shape in profile view			√							Circular
Grain: size of mark of germ						√				Large
Grain: texture of endosperm (in longitudinal section)			√							75 % corneous
Grain: colour of vitreous albumen			√							Yellow
Grain: lustre					√					Medium

[a] *PPVFRA* Protection of Plant Varieties and Farmers Rights Act

22 ICSSH 58 Salient Features

1. Pedigree: ICSA 731 × ICSA 93046 (First A2 *cms* system based hybrid)
2. Days to 50 % flowering: 79 days.
3. Plant height (m): 3.4
4. Plant girth (mm): 15.0
5. Biomass yield (t ha^{-1}): 57.5
6. Juice yield (t ha^{-1}): 29.4
7. Juice extraction (%): 51
8. Brix (%): 17
9. Sugar yield (t ha^{-1}): 3.8
10. Grain yield (t ha^{-1}): 2.9
11. Male fertility restoration (%): 93
12. Sucrose (%): 9.9
13. Glucose (%): 0.3
14. Fructose (%): 1.2
15. pH of juice: 5.0
16. Electrical conductivity of juice (mS m^{-1}) * milli siemens per meter: 9.6
17. Tolerance to: Aphids, Shootfly, Anthracnose.
18. Adaptation: Rainy season

Traits recorded as per guidelines for sorghum as approved by PPVFRA[a]

Characteristics	Characteristic value of candidate hybrid									Remarks measured value etc.
	1	2	3	4	5	6	7	8	9	
Seedling: anthocyanin colouration of coleoptile	✓									Purple
Leaf sheath: anthocyanin colouration		✓								Purple
Leaf: midrib colour (5th fully developed leaf)			✓							Brown
Plant: time of panicle emergence (50 % of the plants with complete panicle emergence)						✓				79
Plant: natural height of foliage up to base of flag leaf						✓				3.4
Flag leaf: yellow colouration of midrib	✓									Absent
Lemma: arista formation (awns)	✓									Absent
Stigma: anthocyanin coloration	✓									Absent
Stigma: yellow coloration	✓									Absent
Stigma length (mm)			✓							Short
Flower with pedicel: length of flower					✓					Medium
Anther: length					✓					Medium
Anther: colour of dry anther				✓						Orange
Glume: color				✓						Red
Plant: total height						✓				3.7
Stem: diameter (at lower one-third height of plant) (mm)						✓				15.0
Leaf: length of blade of the third leaf from top including flag leaf (cm)							✓			85.9
Leaf: width of blade of the third leaf from top including flag leaf (cm)							✓			9.3
Panicle: length without peduncle					✓					20.7
Panicle: length of branches (middle third of panicle)					✓					6.7
Panicle: density at maturity (ear head compactness)						✓				Semi compact
Panicle: shape			✓							Symmetric
Neck of panicle: visible length above sheath (cm)			✓							8.3
Glume coverage (%)			✓							50
Threshability					✓					Partly threshable
Caryopsis: colour after threshing	✓									White
Grain: weight of 1000 grains (g)					✓					25.0
Grain: shape in dorsal view			✓							Circular
Grain: shape in profile view			✓							Circular
Grain: size of mark of germ					✓					Medium
Grain: texture of endosperm (in longitudinal section)						✓				25 % corneous
Grain: colour of vitreous albumen			✓							Yellow
Grain: lustre					✓					Medium

[a] *PPVFRA* Protection of Plant Varieties and Farmers Rights Act

23 ICSSH 25 Salient Features

1. Pedigree: ICSA 675 × ICSV 700
2. Days to 50 % flowering: 77 days.
3. Plant height (m): 3.5
4. Plant girth (mm): 18.1
5. Biomass yield (t ha^{-1}): 54.8
6. Juice yield (t ha^{-1}): 28.5
7. Juice extraction (%): 52
8. Brix (%): 17
9. Sugar yield (t ha^{-1}): 3.6
10. Grain yield (t ha^{-1}): 3.1
11. Male fertility restoration (%): 86
12. Sucrose (%): 6.3
13. Glucose (%): 1.3
14. Fructose (%): 0.3
15. pH of juice: 5.0
16. Electrical conductivity of juice (mS m^{-1}) * milli siemens per meter: 10.3
17. Tolerance to: Aphids, Rust.
18. Adaptation: Rainy season.

Traits recorded as per guidelines for sorghum as approved by PPVFRA[a]

Characteristics	1	2	3	4	5	6	7	8	9	Remarks measured value etc.
Seedling: anthocyanin colouration of coleoptile	√									Purple
Leaf sheath: anthocyanin colouration		√								Purple
Leaf: midrib colour (5th fully developed leaf)			√							Brown
Plant: time of panicle emergence (50 % of the plants with complete panicle emergence)							√			77
Plant: natural height of foliage up to base of flag leaf							√			3.2
Flag leaf: yellow colouration of midrib	√									Absent
Lemma: arista formation (awns)	√									Absent
Stigma: anthocyanin coloration	√									Absent
Stigma: yellow coloration	√									Absent
Stigma length (mm)					√					Medium
Flower with pedicel: length of flower					√					Medium
Anther: length					√					Medium
Anther: colour of dry anther				√						Orange
Glume: color					√					Red
Plant: total height						√				3.5
Stem: diameter (at lower one-third height of plant) (mm)							√			18.1
Leaf: length of blade of the third leaf from top including flag leaf (cm)								√		76.7
Leaf: width of blade of the third leaf from top including flag leaf (cm)								√		9.1
Panicle: length without peduncle					√					22.9
Panicle: length of branches (middle third of panicle)					√					7.8
Panicle: density at maturity (ear head compactness)					√					Semi loose
Panicle: shape			√							Symmetric
Neck of panicle: visible length above sheath (cm)					√					7.6
Glume coverage (%)					√					25
Threshability					√					Partly threshable
Caryopsis: colour after threshing	√									White
Grain: weight of 1000 grains (g)					√					28.0
Grain: shape in dorsal view			√							Circular
Grain: shape in profile view			√							Circular
Grain: size of mark of germ							√			Large
Grain: texture of endosperm (in longitudinal section)							√			25 % corneous
Grain: colour of vitreous albumen			√							Yellow
Grain: lustre							√			Lustrous

[a] *PPVFRA* Protection of Plant Varieties and Farmers Rights Act

24 CSH 22 SS Salient Features

1. Pedigree: ICSA 38 × SSV 84 (Developed by Directorate of Sorghum Research)
2. Days to 50 % flowering: 84 days.
3. Plant height (m): 3.3
4. Plant girth (mm): 18.5
5. Biomass yield (t ha^{-1}): 51.0
6. Juice yield (t ha^{-1}): 22.8
7. Juice extraction (%): 44.3
8. Brix (%): 17
9. Sugar yield (t ha^{-1}): 2.9
10. Grain yield (t ha^{-1}): 2.7
11. Male fertility restoration (%): 86
12. Sucrose (%): 8.5
13. Glucose (%): 0.3
14. Fructose (%): 1.6
15. pH of juice: 5.0
16. Electrical conductivity of juice (mS m^{-1}) * milli siemens per meter: 8.9
17. Tolerance to: Aphids, Shoot fly.
18. Adaptation: Rainy season/Post rainy season.

Traits recorded as per guidelines for sorghum as approved by PPVFRA[a]

Characteristics	Characteristic value of candidate hybrid									Remarks measured value etc.
	1	2	3	4	5	6	7	8	9	
Seedling: anthocyanin colouration of coleoptile		√								Yellow green
Leaf sheath: anthocyanin colouration	√									Yellow green
Leaf: midrib colour (5th fully developed leaf)	√									Yellow green
Plant: time of panicle emergence (50 % of the plants with complete panicle emergence)							√			84
Plant: natural height of foliage up to base of flag leaf							√			3.0
Flag leaf: yellow colouration of midrib	√									Absent
Lemma: arista formation (awns)	√									Absent
Stigma: anthocyanin coloration	√									Present
Stigma: yellow coloration				√						Absent
Stigma length (mm)			√							Medium
Flower with pedicel: length of flower				√						Short
Anther: length				√						Short
Anther: colour of dry anther			√							Grayed orange
Glume: color				√						Grayed purple
Plant: total height						√				3.3
Stem: diameter (at lower one-third height of plant) (mm)						√				Medium
Leaf: length of blade of the third leaf from top including flag leaf (cm)								√		Very long
Leaf: width of blade of the third leaf from top including flag leaf (cm)								√		Vey broad
Panicle: length without peduncle					√					Medium
Panicle: length of branches (middle third of panicle)					√					Medium
Panicle: density at maturity (ear head compactness)					√					Semi compact
Panicle: shape			√							Symmetric
Neck of panicle: visible length above sheath (cm)						√				Absent or very short
Glume coverage (%)	√									Short
Threshability	√									Partly threshable
Caryopsis: colour after threshing	√									Grayed orange
Grain: weight of 1000 grains (g)					√					30.0
Grain: shape in dorsal view			√							Circular
Grain: shape in profile view			√							Elliptic
Grain: size of mark of germ						√				Medium
Grain: texture of endosperm (in longitudinal section)						√				25 % corneous
Grain: colour of vitreous albumen			√							Grayed yellow
Grain: lustre						√				Lustrous

[a] *PPVFRA* Protection of Plant Varieties and Farmers Rights Act

Post-rainy Season Cultivars and Hybrid Parents

P. Srinivasa Rao, C. Ganesh Kumar, Belum V. S. Reddy, A. Kamal, H. C. Sharma and R. P. Thakur

Abstract This chapter entitled "Post-rainy season cultivars and hybrid parents" gives a focussed description of improved sweet sorghum varieties/restorers, hybrids and female hybrid parents (as per PPVFRA). The coloured picture of the line is provided for easy identification. The genetic variability for all the metric traits is conspicuous. However, some of the qualitative traits like anthocyanin coloration, glume color the differences between the lines are indistinct or trivial. Among the biochemical traits such as sucrose, glucose and fructose, the differences are noteworthy among the cultivars. The poor productivity of tropical sorghums during post-rainy season is attributed to photo-thermo sensitivity.

Keywords Hybrids · Varieties, Female hybrid parents, Pedigree · Post-rainy · Rabi · Quantitative traits · Biomass · Panicle · Grain · Glucose · Fructose · Sucrose · Brix % · Shoot fly

P. Srinivasa Rao (✉) · B. V. S. Reddy · H. C. Sharma · R. P. Thakur
International Crops Research Institute for the Semi-Arid Tropics (ICRISAT),
Patancheru 502324, India
e-mail: psrao72@gmail.com

C. Ganesh Kumar · A. Kamal
Chemical Biology Laboratory, CSIR-Indian Institute of Chemical Technology (CSIR-IICT),
Uppal Road, Hyderabad, 500607 India

P. S. Rao and C. G. Kumar (eds.), *Characterization of Improved Sweet Sorghum Cultivars*, SpringerBriefs in Agriculture,
DOI: 10.1007/978-81-322-0783-2_4, © The Author(s) 2013

1 ICSV 700 Salient Features

1. Pedigree: (IS 1082 × SC 108-3)-1-1-1-1
2. Days to 50 % flowering: 81 days
3. Plant height (m): 2.0
4. Plant girth (mm): 13.2
5. Biomass yield (t ha^{-1}): 24.7
6. Juice yield (t ha^{-1}): 6.9
7. Juice extraction (%): 27.7
8. Brix (%): 13.0
9. Sugar yield (t ha^{-1}): 0.7
10. Grain yield (t ha^{-1}): 2.6
11. Sucrose (%): 4.9
12. Glucose (%): 1.7
13. Fructose (%): 1.0
14. pH of juice: 5.3
15. Electrical conductivity of juice (mS m^{-1}—milli siemens per metre): 11.3
16. Tolerance to: Stem borer, Shoot fly, Rust.
17. Adaptation: Post-rainy season

Traits recorded as per guidelines for sorghum as approved by PPVFRA[a]

Characteristics	Characteristic value of candidate variety									Remarks measured value etc.
	1	2	3	4	5	6	7	8	9	
Seedling: anthocyanin colouration of coleoptile		√								Purple
Leaf sheath: anthocyanin colouration			√							Purple
Leaf: midrib colour (5th fully developed leaf)		√								Dull green
Plant: time of panicle emergence (50 % of the plants with complete panicle emergence)							√			81
Plant: natural height of foliage up to base of flag leaf					√					1.8 m
Flag leaf: extension of discoloration of midrib	√									Absent
Flag leaf: intensity of green coloration of midrib compared to the blade		√								Same color
Flag leaf: yellow coloration of midrib	√									Absent
Glume hair color	√									Absent
Lemma: arista formation (awns)									√	Strong
Stigma: anthocyanin coloration	√									Absent
Stigma: yellow coloration					√					Medium
Stigma length (mm)					√					Medium
Flower with pedicel: length of flower					√					Medium
Anther: length					√					Medium
Anther: colour of dry anther				√						Orange
Glume: color					√					Red
Plant: total height						√				2.0 m
Stem: diameter (at lower one-third height of plant) (mm)			√							13.2
Leaf: length of blade of the third leaf from top including flag leaf (cm)					√					54.3
Leaf: width of blade of the third leaf from top including flag leaf (cm)						√				6.9
Panicle: length without peduncle			√							16.6
Panicle: length of branches (middle third of panicle)				√						6.4
Panicle: density at maturity (ear head compactness)						√				Semi compact
Panicle: shape			√							Symmetric
Neck of panicle: visible length above sheath (cm)			√							9.8
Glume coverage (%)					√					67
Shattering			√							Low
Threshability					√					Partly threshable
Grain form		√								Single
Caryopsis: colour after threshing		√								White
Grain: weight of 1000 grains (g)					√					30.0
Grain: shape in dorsal view			√							Circular
Grain: shape in profile view			√							Circular
Grain: size of mark of germ						√				Large
Grain: texture of endosperm (in longitudinal section)					√					50 % corneous
Grain: colour of vitreous albumen			√							Yellow
Grain: lustre					√					Medium
Seedling vigor score		√								Most vigorous
Leaf glossy score		√								Glossy
Plant aspect score			√							Average

[a] *PPVFRA* Protection of Plant Varieties and Farmers Rights Act

2 ICSV 25279 Salient Features

1. Pedigree: (ICSV 93046 × SSV 84)-7-2-1-2
2. Days to 50 % flowering: 83 days
3. Plant height (m): 2.0
4. Plant girth (mm): 15.4
5. Biomass yield (t ha^{-1}): 25.8
6. Juice yield (t ha^{-1}): 9.1
7. Juice extraction (%): 35.1
8. Brix (%): 16
9. Sugar yield (t ha^{-1}): 1.1
10. Grain yield (t ha^{-1}): 3.5
11. Sucrose (%): 7.4
12. Glucose (%): 1.8
13. Fructose (%): 1.2
14. pH of juice: 5.3
15. Electrical conductivity of juice (mS m^{-1}—milli siemens per metre): 12.9
16. Tolerance to: Anthracnose
17. Adaptation: Post-rainy season

Traits recorded as per guidelines for sorghum as approved by PPVFRA[a]

Characteristics	Characteristic value of candidate variety									Remarks measured value etc.
	1	2	3	4	5	6	7	8	9	
Seedling: anthocyanin colouration of coleoptile		√								Purple
Leaf sheath: anthocyanin colouration			√							Purple
Leaf: midrib colour (5th fully developed leaf)		√								Dull green
Plant: time of panicle emergence (50 % of the plants with complete panicle emergence)						√				83
Plant: natural height of foliage up to base of flag leaf				√						1.8 m
Flag leaf: extension of discoloration of midrib					√					Medium
Flag leaf: intensity of green coloration of midrib compared to the blade			√							Same color
Flag leaf: yellow coloration of midrib	√									Absent
Glume hair color	√									Absent
Lemma: arista formation (awns)								√		Strong
Stigma: anthocyanin coloration	√									Absent
Stigma: yellow coloration					√					Medium
Stigma length (mm)					√					Medium
Flower with pedicel: length of flower					√					Medium
Anther: length			√							Short
Anther: colour of dry anther			√							Orange
Glume: color					√					Red
Plant: total height						√				2.0
Stem: diameter (at lower one-third height of plant) (mm)			√							15.4
Leaf: length of blade of the third leaf from top including flag leaf (cm)						√				65
Leaf: width of blade of the third leaf from top including flag leaf (cm)						√				7.8
Panicle: length without peduncle							√			16.2
Panicle: length of branches (middle third of panicle)					√					5.7
Panicle: density at maturity (ear head compactness)						√				Semi compact
Panicle: shape			√							Symmetric
Neck of panicle: visible length above sheath (cm)					√					11.4
Glume coverage (%)				√						48
Shattering			√							Low
Threshability					√					Partly threshable
Grain form		√								Single
Caryopsis: colour after threshing				√						Yellow
Grain: weight of 1000 grains (g)					√					33.0
Grain: shape in dorsal view			√							Circular
Grain: shape in profile view			√							Circular
Grain: size of mark of germ						√				Large
Grain: texture of endosperm (in longitudinal section)						√				25 % corneous
Grain: colour of vitreous albumen				√						Yellow
Grain: lustre		√								Non-lustrous
Seedling vigor score		√								Most vigorous
Leaf glossy score		√								Glossy
Plant aspect score					√					Below average

[a] *PPVFRA* Protection of Plant Varieties and Farmers Rights Act

3 ICSV 25284 Salient Features

1. Pedigree: (DSV 4 × SSV 84)-2-5-1-1
2. Days to 50 % flowering: 67 days
3. Plant height (m): 1.8
4. Plant girth (mm): 14.1
5. Biomass yield (t ha^{-1}): 29.7
6. Juice yield (t ha^{-1}): 7.7
7. Juice extraction (%): 25.7
8. Brix (%): 14
9. Sugar yield (t ha^{-1}): 0.8
10. Grain yield (t ha^{-1}): 2.7
11. Sucrose (%): 6.9
12. Glucose (%): 1.3
13. Fructose (%): 1.7
14. pH of juice: 5.3
15. Electrical conductivity of juice (mS m^{-1}—milli siemens per metre): 14.9
16. Tolerance to: Aphids.
17. Adaptation: Post-rainy season

Traits recorded as per guidelines for sorghum as approved by PPVFRA[a]

Characteristics	Characteristic value of candidate variety									Remarks measured value etc.
	1	2	3	4	5	6	7	8	9	
Seedling: anthocyanin colouration of coleoptile	√									Purple
Leaf sheath: anthocyanin colouration		√								Purple
Leaf: midrib colour (5th fully developed leaf)	√									Dull green
Plant: time of panicle emergence (50 % of the plants with complete panicle emergence)				√						72
Plant: natural height of foliage up to base of flag leaf				√						1.6 m
Flag leaf: extension of discoloration of midrib				√						Medium
Flag leaf: intensity of green coloration of midrib compared to the blade		√								Same color
Flag leaf: yellow coloration of midrib	√									Absent
Glume hair color	√									Absent
Lemma: arista formation (awns)	√									Absent
Stigma: anthocyanin coloration	√									Absent
Stigma: yellow coloration				√						Medium
Stigma length (mm)			√							Short
Flower with pedicel: length of flower			√							Medium
Anther: length				√						Medium
Anther: colour of dry anther			√							Orange
Glume: color							√			Black
Plant: total height				√						1.8 m
Stem: diameter (at lower one-third height of plant) (mm)		√								14.1
Leaf: length of blade of the third leaf from top including flag leaf (cm)					√					68.8
Leaf: width of blade of the third leaf from top including flag leaf (cm)								√		8.7
Panicle: length without peduncle			√							19.1
Panicle: length of branches (middle third of panicle)				√						6.8
Panicle: density at maturity (ear head compactness)				√						Semi loose
Panicle: shape			√							Symmetric
Neck of panicle: visible length above sheath (cm)			√							12.2
Glume coverage (%)	√									25
Shattering			√							Low
Threshability				√						Partly threshable
Grain form		√								Single
Caryopsis: colour after threshing	√									White
Grain: weight of 1000 grains (g)					√					41.0
Grain: shape in dorsal view			√							Circular
Grain: shape in profile view			√							Circular
Grain: size of mark of germ					√					Large
Grain: texture of endosperm (in longitudinal section)				√						50 % corneous
Grain: colour of vitreous albumen			√							Yellow
Grain: lustre				√						Medium
Seedling vigor score	√									Most vigorous
Leaf glossy score	√									Glossy
Plant aspect score			√							Below average

[a] *PPVFRA* Protection of Plant Varieties and Farmers Rights Act

4 ICSV 93046 Salient Features

1. Pedigree: [(((IS 1082 × SC 108 -3)-1-1-1-1-1) x (((IS 5622 × CS 3541)-20-1-1-1-1-1-1 x (UChV2 × Bulk Y-55)-1-5-1)]-9-3-1-1-1
2. Days to 50 % flowering: 83 days
3. Plant height (m): 2.0
4. Plant girth (mm): 15.3
5. Biomass yield (t ha^{-1}): 33.9
6. Juice yield (t ha^{-1}): 11.2
7. Juice extraction (%): 32.9
8. Brix (%): 16
9. Sugar yield (t ha^{-1}): 1.3
10. Grain yield (t ha^{-1}): 3.1
11. Sucrose (%): 5.6
12. Glucose (%): 1.0
13. Fructose (%): 1.7
14. pH of juice: 5.3
15. Electrical conductivity of juice (mS m^{-1}—milli siemens per metre): 10.5
16. Tolerance to: Shoot fly, Anthracnose, Grain mold and Downy mildew.
17. Adaptation: Rainy season/Post-rainy season.

Traits recorded as per guidelines for sorghum as approved by PPVFRA[a]

Characteristics	Characteristic value of candidate variety									Remarks measured value etc.
	1	2	3	4	5	6	7	8	9	
Seedling: anthocyanin colouration of coleoptile	✓									Purple
Leaf sheath: anthocyanin colouration		✓								Purple
Leaf: midrib colour (5th fully developed leaf)	✓									Dull green
Plant: time of panicle emergence (50 % of the plants with complete panicle emergence)						✓				83
Plant: natural height of foliage up to base of flag leaf				✓						1.9 m
Flag leaf: extension of discoloration of midrib	✓									Absent
Flag leaf: intensity of green coloration of midrib compared to the blade	✓									Same color
Flag leaf: yellow coloration of midrib	✓									Absent
Glume hair color	✓									Absent
Lemma: arista formation (awns)							✓			Strong
Stigma: anthocyanin coloration	✓									Absent
Stigma: yellow coloration				✓						Medium
Stigma length (mm)				✓						Medium
Flower with pedicel: length of flower			✓							Short
Anther: length			✓							Short
Anther: colour of dry anther				✓						Orange
Glume: color		✓								Straw
Plant: total height						✓				2.0 m
Stem: diameter (at lower one-third height of plant) (mm)			✓							15.3
Leaf: length of blade of the third leaf from top including flag leaf (cm)				✓						52.6
Leaf: width of blade of the third leaf from top including flag leaf (cm)						✓				7
Panicle: length without peduncle			✓							15.1
Panicle: length of branches (middle third of panicle)				✓						6.4
Panicle: density at maturity (ear head compactness)						✓				Semi compact
Panicle: shape			✓							Symmetric
Neck of panicle: visible length above sheath (cm)				✓						10.3
Glume coverage (%)			✓							42
Shattering			✓							Low
Threshability				✓						Partly threshable
Grain form	✓									Single
Caryopsis: colour after threshing	✓									White
Grain: weight of 1000 grains (g)				✓						33.0
Grain: shape in dorsal view			✓							Circular
Grain: shape in profile view			✓							Circular
Grain: size of mark of germ				✓						Medium
Grain: texture of endosperm (in longitudinal section)						✓				25 % corneous
Grain: colour of vitreous albumen			✓							Yellow
Grain: lustre				✓						Medium
Seedling vigor score	✓									Most vigorous
Leaf glossy score	✓									Glossy
Plant aspect score			✓							Average

[a] *PPVFRA* Protection of Plant Varieties and Farmers Rights Act

5 SSV 74 Salient Features

1. Pedigree: Selection from PAB 74 (Sudan; Bred by UAS, Dharwad)
2. Days to 50 % flowering: 73 days
3. Plant height (m): 1.7
4. Plant girth (mm): 12.8
5. Biomass yield (t ha^{-1}): 23.2
6. Juice yield (t ha^{-1}): 8.4
7. Juice extraction (%): 36
8. Brix (%): 13
9. Sugar yield (t ha^{-1}): 0.8
10. Grain yield (t ha^{-1}): 2.7
11. Sucrose (%): 5.1
12. Glucose (%): 1.5
13. Fructose (%): 1.1
14. pH of juice: 5.2
15. Electrical conductivity of juice (mS m^{-1}—milli siemens per metre): 12.9
16. Tolerance to: Aphids, Rust.
17. Adaptation: Post-rainy season

Traits recorded as per guidelines for sorghum as approved by PPVFRA[a]

Characteristics	Characteristic value of candidate variety									Remarks measured value etc.
	1	2	3	4	5	6	7	8	9	
Seedling: anthocyanin colouration of coleoptile		✓								Purple
Leaf sheath: anthocyanin colouration	✓									Tan
Leaf: midrib colour (5th fully developed leaf)		✓								Dull green
Plant: time of panicle emergence (50 % of the plants with complete panicle emergence)					✓					73
Plant: natural height of foliage up to base of flag leaf			✓							1.5 m
Flag leaf: extension of discoloration of midrib	✓									Absent
Flag leaf: intensity of green coloration of midrib compared to the blade		✓								Same color
Flag leaf: yellow coloration of midrib	✓									Absent
Glume hair color	✓									Absent
Lemma: arista formation (awns)	✓									Absent
Stigma: anthocyanin coloration	✓									Absent
Stigma: yellow coloration					✓					Medium
Stigma length (mm)			✓							Short
Flower with pedicel: length of flower					✓					Medium
Anther: length					✓					Medium
Anther: colour of dry anther				✓						Orange
Glume: color								✓		Black
Plant: total height					✓					1.7 m
Stem: diameter (at lower one-third height of plant) (mm)			✓							12.8
Leaf: length of blade of the third leaf from top including flag leaf (cm)							✓			65
Leaf: width of blade of the third leaf from top including flag leaf (cm)							✓			7.9
Panicle: length without peduncle			✓							18.8
Panicle: length of branches (middle third of panicle)					✓					7.1
Panicle: density at maturity (ear head compactness)					✓					Semi loose
Panicle: shape			✓							Symmetric
Neck of panicle: visible length above sheath (cm)			✓							8.7
Glume coverage (%)	✓									25
Shattering			✓							Low
Threshability					✓					Partly threshable
Grain form	✓									Single
Caryopsis: colour after threshing	✓									White
Grain: weight of 1000 grains (g)							✓			43.0
Grain: shape in dorsal view			✓							Circular
Grain: shape in profile view			✓							Circular
Grain: size of mark of germ						✓				Large
Grain: texture of endosperm (in longitudinal section)						✓				25 % corneous
Grain: colour of vitreous albumen			✓							Yellow
Grain: lustre					✓					Medium
Seedling vigor score	✓									Most vigorous
Leaf glossy score	✓									Glossy
Plant aspect score			✓							Average

[a] *PPVFRA* Protection of Plant Varieties and Farmers Rights Act

6 SSV 84 Salient Features

1. Pedigree: Selection from IS 23568 (Bred by MPKV, Rahuri)
2. Days to 50 % flowering: 77 days
3. Plant height (m): 1.5
4. Plant girth (mm): 13.7
5. Biomass yield (t ha^{-1}): 20.9
6. Juice yield (t ha^{-1}): 9.1
7. Juice extraction (%): 43.4
8. Brix (%): 1.0
9. Sugar yield (t ha^{-1}): 0.7
10. Grain yield (t ha^{-1}): 3.0
11. Sucrose (%): 3.7
12. Glucose (%): 1.2
13. Fructose (%): 1.1
14. pH of juice: 5.2
15. Electrical conductivity of juice (mS m^{-1}—milli siemens per metre): 16.6
16. Tolerance to: Aphids, Shoot fly.
17. Adaptation: Post-rainy season

Traits recorded as per guidelines for sorghum as approved by PPVFRA[a]

Characteristics	1	2	3	4	5	6	7	8	9	Remarks measured value etc.
Seedling: anthocyanin colouration of coleoptile		√								Purple
Leaf sheath: anthocyanin colouration			√							Purple
Leaf: midrib colour (5th fully developed leaf)		√								Dull green
Plant: time of panicle emergence (50 % of the plants with complete panicle emergence)					√					72
Plant: natural height of foliage up to base of flag leaf			√							1.3 m
Flag leaf: extension of discoloration of midrib							√			Strong
Flag leaf: intensity of green coloration of midrib compared to the blade		√								Same color
Flag leaf: yellow coloration of midrib	√									Absent
Glume hair color	√									Absent
Lemma: arista formation (awns)	√									Absent
Stigma: anthocyanin coloration	√									Absent
Stigma: yellow coloration					√					Medium
Stigma length (mm)			√							Short
Flower with pedicel: length of flower					√					Medium
Anther: length			√							Short
Anther: colour of dry anther				√						Orange
Glume: color					√					Red
Plant: total height					√					1.5 m
Stem: diameter (at lower one-third height of plant) (mm)			√							13.7
Leaf: length of blade of the third leaf from top including flag leaf (cm)						√				78.4
Leaf: width of blade of the third leaf from top including flag leaf (cm)									√	8.1
Panicle: length without peduncle			√							16.1
Panicle: length of branches (middle third of panicle)					√					6.1
Panicle: density at maturity (ear head compactness)					√					Semi loose
Panicle: shape			√							Symmetric
Neck of panicle: visible length above sheath (cm)					√					10.2
Glume coverage (%)	√									25
Shattering			√							Low
Threshability					√					Partly threshable
Grain form	√									Single
Caryopsis: colour after threshing	√									White
Grain: weight of 1000 grains (g)					√					32.0
Grain: shape in dorsal view			√							Circular
Grain: shape in profile view			√							Circular
Grain: size of mark of germ							√			Large
Grain: texture of endosperm (in longitudinal section)					√					50 % corneous
Grain: colour of vitreous albumen			√							Yellow
Grain: lustre					√					Medium
Seedling vigor score	√									Most vigorous
Leaf glossy score	√									Glossy
Plant aspect score				√						Below average

[a] *PPVFRA* Protection of Plant Varieties and Farmers Rights Act

7 RSSV 9 (CSV 19SS) Salient Features

1. Pedigree: (RSSV 2 × SPV 462)-2-1-1 (Bred by MPKV, Rahuri)
2. Days to 50 % flowering: 71 days
3. Plant height (m): 1.9
4. Plant girth (mm): 14.4
5. Biomass yield (t ha^{-1}): 27.5
6. Juice yield (t ha^{-1}): 10.6
7. Juice extraction (%): 38.7
8. Brix (%): 14
9. Sugar yield (t ha^{-1}): 1.1
10. Grain yield (t ha^{-1}): 3.1
11. Sucrose (%): 4.1
12. Glucose (%): 1.5
13. Fructose (%): 1.2
14. pH of juice: 5.2
15. Electrical conductivity of juice (mS m^{-1}—milli siemens per metre): 16.0
16. Tolerance to: Aphids, Shoot fly
17. Adaptation: Post-rainy season

Traits recorded as per guidelines for sorghum as approved by PPVFRA[a]

Characteristics	Characteristic value of candidate variety									Remarks measured value etc.
	1	2	3	4	5	6	7	8	9	
Seedling: anthocyanin colouration of coleoptile	√									Purple
Leaf sheath: anthocyanin colouration		√								Purple
Leaf: midrib colour (5th fully developed leaf)	√									Dull green
Plant: time of panicle emergence (50 % of the plants with complete panicle emergence)	√									71
Plant: natural height of foliage up to base of flag leaf					√					1.8 m
Flag leaf: extension of discoloration of midrib					√					Medium
Flag leaf: intensity of green coloration of midrib compared to the blade			√							Same color
Flag leaf: yellow coloration of midrib	√									Absent
Glume hair color	√									Absent
Lemma: arista formation (awns)								√		Strong
Stigma: anthocyanin coloration	√									Absent
Stigma: yellow coloration					√					Medium
Stigma length (mm)					√					Medium
Flower with pedicel: length of flower					√					Medium
Anther: length					√					Medium
Anther: colour of dry anther				√						Orange
Glume: color					√					Red
Plant: total height							√			1.9
Stem: diameter (at lower one-third height of plant) (mm)			√							14.4
Leaf: length of blade of the third leaf from top including flag leaf (cm)						√				63.9
Leaf: width of blade of the third leaf from top including flag leaf (cm)						√				7.3
Panicle: length without peduncle			√							13.7
Panicle: length of branches (middle third of panicle)				√						5.7
Panicle: density at maturity (ear head compactness)						√				Semi compact
Panicle: shape			√							Symmetric
Neck of panicle: visible length above sheath (cm)	√									2.0
Glume coverage (%)			√							33
Shattering			√							Low
Threshability				√						Partly threshable
Grain form		√								Single
Caryopsis: colour after threshing			√							Yellow
Grain: weight of 1000 grains (g)			√							33.0
Grain: shape in dorsal view			√							Circular
Grain: shape in profile view			√							Circular
Grain: size of mark of germ						√				Large
Grain: texture of endosperm (in longitudinal section)					√					50 % corneous
Grain: colour of vitreous albumen			√							Yellow
Grain: lustre				√						Medium
Seedling vigor score		√								Most vigorous
Leaf glossy score		√								Glossy
Plant aspect score			√							Good

[a] *PPVFRA* Protection of Plant Varieties and Farmers Rights Act

8 ICSB 38 Salient Features

1. Pedigree: [(BT × 623 × MR 862)B lines bulk]-5-1-3-5
2. Days to 50 % flowering: 69 days
3. Plant height (m): 0.9
4. Plant girth (mm): 16.0
5. Biomass yield (t ha^{-1}): 9.6
6. Juice yield (t ha^{-1}): 3.3
7. Juice extraction (%): 34.1
8. Brix (%): 11
9. Sugar yield (t ha^{-1}): 0.3
10. Grain yield (t ha^{-1}): 3.8
11. Sucrose (%): 5.0
12. Glucose (%): 2.0
13. Fructose (%): 1.5
14. pH of juice: 5.3
15. Electrical conductivity of juice (mS m^{-1}—milli siemens per metre): 16.9
16. Tolerance to: Shoot fly, Stem borer, Rust
17. Adaptation: Post-rainy season

Traits recorded as per guidelines for sorghum as approved by PPVFRA[a]

Characteristics	1	2	3	4	5	6	7	8	9	Remarks measured value etc.
Seedling: anthocyanin colouration of coleoptile		✓								Purple
Leaf sheath: anthocyanin colouration	✓									Tan
Leaf: midrib colour (5th fully developed leaf)		✓								Dull green
Plant: time of panicle emergence (50 % of the plants with complete panicle emergence)					✓					69
Plant: natural height of foliage up to base of flag leaf			✓							0.7 m
Flag leaf: extension of discoloration of midrib					✓					Medium
Flag leaf: intensity of green coloration of midrib compared to the blade		✓								Same color
Flag leaf: yellow coloration of midrib	✓									Absent
Glume hair color	✓									Absent
Lemma: arista formation (awns)	✓									Absent
Stigma: anthocyanin coloration	✓									Absent
Stigma: yellow coloration					✓					Medium
Stigma length (mm)			✓							Short
Flower with pedicel: length of flower			✓							Short
Anther: length			✓							Short
Anther: colour of dry anther				✓						Orange
Glume: color					✓					Red
Plant: total height			✓							0.9
Stem: diameter (at lower one-third height of plant) (mm)			✓							16.0
Leaf: length of blade of the third leaf from top including flag leaf (cm)							✓			70.3
Leaf: width of blade of the third leaf from top including flag leaf (cm)							✓			7.2
Panicle: length without peduncle					✓					27.7
Panicle: length of branches (middle third of panicle)					✓					9.2
Panicle: density at maturity (ear head compactness)					✓					Semi loose
Panicle: shape			✓							Symmetric
Neck of panicle: visible length above sheath (cm)								✓		20.3
Glume coverage (%)	✓									25
Shattering			✓							Low
Threshability					✓					Partly threshable
Grain form	✓									Single
Caryopsis: colour after threshing	✓									White
Grain: weight of 1000 grains (g)					✓					28.0
Grain: shape in dorsal view			✓							Circular
Grain: shape in profile view			✓							Circular
Grain: size of mark of germ						✓				Large
Grain: texture of endosperm (in longitudinal section)							✓			25 % corneous
Grain: colour of vitreous albumen			✓							Yellow
Grain: lustre					✓					Medium
Seedling vigor score	✓									Most vigorous
Leaf glossy score	✓									Glossy
Plant aspect score				✓						Below average

[a] *PPVFRA* Protection of Plant Varieties and Farmers Rights Act

9 ICSB 474 Salient Features

1. Pedigree: (IS 18342 × ICSB 6) 11-1-1-2-2
2. Days to 50 % flowering: 74 days
3. Plant height (m): 1.0
4. Plant girth (mm): 12.6
5. Biomass yield (t ha^{-1}): 17.0
6. Juice yield (t ha^{-1}): 4.2
7. Juice extraction (%): 24.4
8. Brix (%): 11
9. Sugar yield (t ha^{-1}): 0.3
10. Grain yield (t ha^{-1}): 3.3
11. Sucrose (%): 4.7
12. Glucose (%): 1.2
13. Fructose (%): 1.0
14. pH of juice: 5.2
15. Electrical conductivity of juice (mS m^{-1}—milli siemens per metre): 17.4
16. Tolerance to: Stem borer
17. Adaptation: Post-rainy season

Traits recorded as per guidelines for sorghum as approved by PPVFRA[a]

Characteristics	Characteristic value of candidate variety									Remarks measured value etc.
	1	2	3	4	5	6	7	8	9	
Seedling: anthocyanin colouration of coleoptile	√									Purple
Leaf sheath: anthocyanin colouration		√								Purple
Leaf: midrib colour (5th fully developed leaf)	√									Dull green
Plant: time of panicle emergence (50 % of the plants with complete panicle emergence)				√						74
Plant: natural height of foliage up to base of flag leaf			√							1.4 m
Flag leaf: extension of discoloration of midrib	√									Absent
Flag leaf: intensity of green coloration of midrib compared to the blade		√								Same color
Flag leaf: yellow coloration of midrib	√									Absent
Glume hair color	√									Absent
Lemma: arista formation (awns)								√		Strong
Stigma: anthocyanin coloration	√									Absent
Stigma: yellow coloration					√					Medium
Stigma length (mm)					√					Medium
Flower with pedicel: length of flower			√							Short
Anther: length			√							Short
Anther: colour of dry anther				√						Orange
Glume: color		√								Straw
Plant: total height					√					1.0 m
Stem: diameter (at lower one-third height of plant) (mm)			√							12.6
Leaf: length of blade of the third leaf from top including flag leaf (cm)							√			66.7
Leaf: width of blade of the third leaf from top including flag leaf (cm)							√			7.6
Panicle: length without peduncle			√							19.4
Panicle: length of branches (middle third of panicle)				√						6.7
Panicle: density at maturity (ear head compactness)					√					Semi compact
Panicle: shape			√							Symmetric
Neck of panicle: visible length above sheath (cm)			√							9.4
Glume coverage (%)					√					67
Shattering			√							Low
Threshability				√						Partly threshable
Grain form		√								Single
Caryopsis: colour after threshing				√						Yellow
Grain: weight of 1000 grains (g)					√					27.0
Grain: shape in dorsal view			√							Circular
Grain: shape in profile view			√							Circular
Grain: size of mark of germ						√				Large
Grain: texture of endosperm (in longitudinal section)						√				50 % corneous
Grain: colour of vitreous albumen			√							Yellow
Grain: lustre	√									Non-lustrous
Seedling vigor score	√									Most vigorous
Leaf glossy score	√									Glossy
Plant aspect score			√							Average

[a] *PPVFRA* Protection of Plant Varieties and Farmers Rights Act

10 ICSB 502 Salient Features

1. Pedigree: [(ICSB 11 × PM 17500-2-1) × PM 17467B]5-2-2
2. Days to 50 % flowering: 78 days
3. Plant height (m): 0.9
4. Plant girth (mm): 16.4
5. Biomass yield (t ha^{-1}): 15.7
6. Juice yield (t ha^{-1}): 3.9
7. Juice extraction (%): 24.8
8. Brix (%): 13
9. Sugar yield (t ha^{-1}): 0.4
10. Grain yield (t ha^{-1}): 2.5
11. Sucrose (%): 4.1
12. Glucose (%): 2.0
13. Fructose (%): 0.9
14. pH of juice: 5.4
15. Electrical conductivity of Juice (mS m^{-1}—milli siemens per metre): 13.5
16. Tolerance to: Midge
17. Adaptation: Post-rainy season

Traits recorded as per guidelines for sorghum as approved by PPVFRA[a]

Characteristics	| 1	2	3	4	5	6	7	8	9	Remarks measured value etc.
Seedling: anthocyanin colouration of coleoptile			√							Purple
Leaf sheath: anthocyanin colouration	√									Tan
Leaf: midrib colour (5th fully developed leaf)			√							Dull green
Plant: time of panicle emergence (50 % of the plants with complete panicle emergence)							√			78
Plant: natural height of foliage up to base of flag leaf			√							0.8 m
Flag leaf: Extension of discoloration of midrib					√					Medium
Flag leaf: intensity of green coloration of midrib compared to the blade		√								Same color
Flag leaf: yellow coloration of midrib	√									Absent
Glume hair color	√									Absent
Lemma: arista formation (awns)	√									Absent
Stigma: anthocyanin coloration	√									Absent
Stigma: yellow coloration						√				Medium
Stigma length (mm)				√						Short
Flower with pedicel: length of flower						√				Medium
Anther: length	√									Short
Anther: colour of dry anther				√						Orange
Glume: color						√				Red
Plant: total height			√							0.9
Stem: diameter (at lower one-third height of plant) (mm)			√							16.4
Leaf: length of blade of the third leaf from top including flag leaf (cm)								√		74.6
Leaf: width of blade of the third leaf from top including flag leaf (cm)								√		7.2
Panicle: length without peduncle						√				24.9
Panicle: length of branches (middle third of panicle)						√				5.8
Panicle: density at maturity (ear head compactness)							√			Semi compact
Panicle: shape			√							Symmetric
Neck of panicle: visible length above sheath (cm)			√							8
Glume coverage (%)	√									25
Shattering			√							Low
Threshability						√				Partly threshable
Grain form	√									Single
Caryopsis: colour after threshing				√						Yellow
Grain: weight of 1000 grains (g)						√				31.0
Grain: shape in dorsal view			√							Circular
Grain: shape in profile view			√							Circular
Grain: size of mark of germ						√				Medium
Grain: texture of endosperm (in longitudinal section)						√				50 % corneous
Grain: colour of vitreous albumen			√							Yellow
Grain: lustre						√				Medium
Seedling vigor score	√									Most vigorous
Leaf glossy score	√									Glossy
Plant aspect score			√							Average

[a] *PPVFRA* Protection of Plant Varieties and Farmers Rights Act

11 ICSB 675 Salient Features

1. Pedigree: (E 36-1 × ICSB 17)12-2
2. Days to 50 % flowering: 76 days
3. Plant height (m): 0.8
4. Plant girth (mm): 19.7
5. Biomass yield (t ha^{-1}): 13.7
6. Juice yield (t ha^{-1}): 4.8
7. Juice extraction (%): 34.8
8. Brix (%): 9
9. Sugar yield (t ha^{-1}): 0.3
10. Grain yield (t ha^{-1}): 3.5
11. Sucrose (%): 4.4
12. Glucose (%): 1.0
13. Fructose (%): 1.5
14. pH of juice: 5.3
15. Electrical conductivity of juice (mS m^{-1}—milli siemens per metre): 18.8
16. Tolerance to: Aphids, Terminal drought
17. Adaptation: Post-rainy season

Traits recorded as per guidelines for sorghum as approved by PPVFRA[a]

Characteristics	1	2	3	4	5	6	7	8	9	Remarks measured value etc.
Seedling: anthocyanin colouration of coleoptile	√									Purple
Leaf sheath: anthocyanin colouration		√								Purple
Leaf: midrib colour (5th fully developed leaf)	√									Dull green
Plant: time of panicle emergence (50 % of the plants with complete panicle emergence)					√					76
Plant: natural height of foliage up to base of flag leaf		√								0.7 m
Flag leaf: extension of discoloration of midrib						√				Strong
Flag leaf: intensity of green coloration of midrib compared to the blade	√									Same color
Flag leaf: yellow coloration of midrib	√									Absent
Glume hair color	√									Absent
Lemma: arista formation (awns)	√									Absent
Stigma: anthocyanin coloration	√									Absent
Stigma: yellow coloration				√						Medium
Stigma length (mm)			√							Short
Flower with pedicel: length of flower				√						Medium
Anther: length				√						Medium
Anther: colour of dry anther			√							Orange
Glume: color							√			Black
Plant: total height		√								0.8
Stem: diameter (at lower one-third height of plant) (mm)		√								19.7
Leaf: length of blade of the third leaf from top including flag leaf (cm)					√					72.1
Leaf: width of blade of the third leaf from top including flag leaf (cm)							√			8.1
Panicle: length without peduncle				√						24.9
Panicle: length of branches (middle third of panicle)					√					8.7
Panicle: density at maturity (ear head compactness)				√						Semi loose
Panicle: shape			√							Symmetric
Neck of panicle: visible length above sheath (cm)	√									3.3
Glume coverage (%)	√									25
Shattering			√							Low
Threshability				√						Partly threshable
Grain form	√									Single
Caryopsis: colour after threshing			√							Yellow
Grain: weight of 1000 grains (g)				√						35.0
Grain: shape in dorsal view		√								Circular
Grain: shape in profile view		√								Circular
Grain: size of mark of germ					√					Large
Grain: texture of endosperm (in longitudinal section)				√						50 % corneous
Grain: colour of vitreous albumen			√							Yellow
Grain: lustre				√						Medium
Seedling vigor score	√									Most vigorous
Leaf glossy score	√									Glossy
Plant aspect score		√								Average

[a] *PPVFRA* Protection of Plant Varieties and Farmers Rights Act

12 ICSB 731 Salient Features

1. Pedigree: ICSV 1171BF
2. Days to 50 % flowering: 69 days
3. Plant height (m): 1.5
4. Plant girth (mm): 16.4
5. Biomass yield (t ha^{-1}): 17.0
6. Juice yield (%): 5.9
7. Juice extraction (%): 34.8
8. Brix (%): 10
9. Sugar yield (t ha^{-1}): 0.4
10. Grain yield (t ha^{-1}): 2.5
11. Sucrose (%): 4.6
12. Glucose (%): 1.2
13. Fructose (%): 1.5
14. pH of juice: 5.3
15. Electrical conductivity of juice (mS m^{-1}—milli siemens per metre): 16.8
16. Tolerance: Anthracnose
17. Adaptation: Post-rainy season

Traits recorded as per guidelines for sorghum as approved by PPVFRA[a]

Characteristics	Characteristic value of candidate variety									Remarks measured value etc.
	1	2	3	4	5	6	7	8	9	
Seedling: anthocyanin colouration of coleoptile	√									Purple
Leaf sheath: anthocyanin colouration		√								Purple
Leaf: midrib colour (5th fully developed leaf)	√									Dull green
Plant: time of panicle emergence (50 % of the plants with complete panicle emergence)					√					69
Plant: natural height of foliage up to base of flag leaf		√								1.3 m
Flag leaf: extension of discoloration of midrib	√									Absent
Flag leaf: intensity of green coloration of midrib compared to the blade	√									Same color
Flag leaf: yellow coloration of midrib	√									Absent
Glume hair color	√									Absent
Lemma: arista formation (awns)	√									Absent
Stigma: anthocyanin coloration	√									Absent
Stigma: yellow coloration				√						Medium
Stigma length (mm)			√							Short
Flower with pedicel: length of flower			√							Short
Anther: length			√							Short
Anther: colour of dry anther				√						Orange
Glume: color		√								Straw
Plant: total height					√					1.5 m
Stem: diameter (at lower one-third height of plant) (mm)			√							16.4
Leaf: length of blade of the third leaf from top including flag leaf (cm)						√				68.2
Leaf: width of blade of the third leaf from top including flag leaf (cm)						√				7.6
Panicle: length without peduncle			√							19.4
Panicle: length of branches (middle third of panicle)				√						6.7
Panicle: density at maturity (ear head compactness)					√					Semi compact
Panicle: shape			√							Symmetric
Neck of panicle: visible length above sheath (cm)			√							6.8
Glume coverage (%)		√								25
Shattering			√							Low
Threshability				√						Partly threshable
Grain form		√								Single
Caryopsis: colour after threshing				√						Yellow
Grain: weight of 1000 grains (g)				√						31.0
Grain: shape in dorsal view			√							Circular
Grain: shape in profile view			√							Circular
Grain: size of mark of germ						√				Large
Grain: texture of endosperm (in longitudinal section)						√				25 % corneous
Grain: colour of vitreous albumen			√							Yellow
Grain: lustre						√				Lustorus
Seedling vigor score		√								Most vigorous
Leaf glossy score		√								Glossy
Plant aspect score				√						Below average

[a] *PPVFRA* Protection of Plant Varieties and Farmers Rights Act

13 ICSSH 25 Salient Features

1. Pedigree: ICSA 675 × ICSV 700
2. Days to 50 % flowering: 77 days
3. Plant height (m): 1.9
4. Plant girth (mm): 14.4
5. Biomass yield (t ha^{-1}): 31.7
6. Juice yield (t ha^{-1}): 13.3
7. Juice extraction (%): 41.9
8. Brix (%): 14
9. Sugar yield (t ha^{-1}): 1.4
10. Grain yield (t ha^{-1}): 2.7
11. Male fertility restoration (%): 94
12. Sucrose (%): 6.8
13. Glucose (%): 1.3
14. Fructose (%): 1.2
15. pH of juice: 5.2
16. Electrical conductivity of juice (mS m^{-1}—milli siemens per metre): 13.0
17. Tolerance to: Aphids, Rust
18. Adaptation: Post-rainy season

Traits recorded as per guidelines for sorghum as approved by PPVFRA[*]

Characteristics	Characteristic value of candidate hybrid									Remarks measured value etc.
	1	2	3	4	5	6	7	8	9	
Seedling: anthocyanin colouration of coleoptile	√									Purple
Leaf sheath: anthocyanin colouration		√								Purple
Leaf: midrib colour (5th fully developed leaf)	√									Dull green
Plant: time of panicle emergence (50 % of the plants with complete panicle emergence)								√		77
Plant: natural height of foliage up to base of flag leaf			√							1.7 m
Flag leaf: extension of discoloration of midrib	√									Absent
Flag leaf: intensity of green coloration of midrib compared to the blade		√								Same color
Flag leaf: yellow coloration of midrib	√									Absent
Glume hair color	√									Absent
Lemma: arista formation (awns)	√									Absent
Stigma: anthocyanin coloration	√									Absent
Stigma: yellow coloration						√				Medium
Stigma length (mm)						√				Medium
Flower with pedicel: length of flower						√				Medium
Anther: length						√				Medium
Anther: colour of dry anther					√					Orange
Glume: color						√				Red
Plant: total height						√				1.9 m
Stem: diameter (at lower one-third height of plant) (mm)			√							14.4
Leaf: length of blade of the third leaf from top including flag leaf (cm)								√		63.78
Leaf: width of blade of the third leaf from top including flag leaf (cm)								√		7.33
Panicle: length without peduncle							√			22
Panicle: length of branches (middle third of panicle)							√			8.5
Panicle: density at maturity (ear head compactness)							√			Semi loose
Panicle: shape			√							Symmetric
Neck of panicle: visible length above sheath (cm)		√								18.78
Glume coverage (%)		√								33.33
Shattering		√								Low
Threshability						√				Partly threshable
Grain form	√									Single
Caryopsis: colour after threshing					√					Yellow
Grain: weight of 1000 grains (g)						√				32.0
Grain: shape in dorsal view			√							Circular
Grain: shape in profile view			√							Circular
Grain: size of mark of germ						√				Medium
Grain: texture of endosperm (in longitudinal section)						√				50 % corneous
Grain: colour of vitreous albumen					√					Yellow
Grain: lustre	√									Non-lustrous
Seedling vigor score	√									Most vigorous
Leaf glossy score	√									Glossy
Plant aspect score		√								Good

[a] PPVFRA Protection of Plant Varieties and Farmers Rights Act

14 ICSSH 28 Salient Features

1. Pedigree: ICSV 474 × SSV 74
2. Days to 50 % flowering: 73 days
3. Plant height (m): 2.1
4. Plant Girth (mm): 12.1
5. Biomass yield (t ha^{-1}): 31.7
6. Juice yield (t ha^{-1}): 11.9
7. Juice extraction (%): 37.5
8. Brix (%): 15
9. Sugar yield (t ha^{-1}): 1.3
10. Grain yield (t ha^{-1}): 2.6
11. Male fertility restoration (%): 91
12. Sucrose (%): 7.9
13. Glucose (%): 1.0
14. Fructose (%): 1.1
15. pH of juice: 5.2
16. Electrical conductivity of juice (mS m^{-1}—milli siemens per metre): 15.4
17. Tolerance to: Aphids, Shoot fly
18. Adaptation: Post-rainy season

Traits recorded as per guidelines for sorghum as approved by PPVFRA[a]

Characteristics	Characteristic value of candidate hybrid									Remarks measured value etc.
	1	2	3	4	5	6	7	8	9	
Seedling: anthocyanin colouration of coleoptile		✓								Purple
Leaf sheath: anthocyanin colouration			✓							Purple
Leaf: midrib colour (5th fully developed leaf)		✓								Dull green
Plant: time of panicle emergence (50 % of the plants with complete panicle emergence)					✓					73
Plant: natural height of foliage up to base of flag leaf			✓							1.9 m
Flag leaf: extension of discoloration of midrib	✓									Absent
Flag leaf: intensity of green coloration of midrib compared to the blade	✓									Same color
Flag leaf: yellow coloration of midrib	✓									Absent
Glume hair color	✓									Absent
Lemma: arista formation (awns)	✓									Absent
Stigma: anthocyanin coloration	✓									Absent
Stigma: yellow coloration					✓					Medium
Stigma length (mm)					✓					Medium
Flower with pedicel: length of flower					✓					Short
Anther: length					✓					Medium
Anther: colour of dry anther				✓						Orange
Glume: color				✓						Light red
Plant: total height					✓					2.1 m
Stem: diameter (at lower one-third height of plant) (mm)			✓							12.1
Leaf: length of blade of the third leaf from top including flag leaf (cm)							✓			65.11
Leaf: width of blade of the third leaf from top including flag leaf (cm)								✓		8.22
Panicle: length without peduncle								✓		43
Panicle: length of branches (middle third of panicle)					✓					7.78
Panicle: density at maturity (ear head compactness)			✓							Loose
Panicle: shape			✓							Symmetric
Neck of panicle: visible length above sheath (cm)						✓				15.33
Glume coverage (%)	✓									25
Shattering		✓								Low
Threshability				✓						Partly threshable
Grain form	✓									Single
Caryopsis: colour after threshing	✓									White
Grain: weight of 1000 grains (g)				✓						12.1
Grain: shape in dorsal view			✓							Circular
Grain: shape in profile view			✓							Circular
Grain: size of mark of germ				✓						Medium
Grain: texture of endosperm (in longitudinal section)						✓				25 % corneous
Grain: colour of vitreous albumen			✓							Yellow
Grain: lustre				✓						Medium
Seedling vigor score	✓									Most vigorous
Leaf glossy score	✓									Glossy
Plant aspect score			✓							Average

[a] *PPVFRA* Protection of Plant Varieties and Farmers Rights Act

15 ICSSH 58 Salient Features

1. Pedigree: ICSA 731 × ICSA 93046 (First A2 *cms* system based hybrid)
2. Days to 50 % flowering: 78 days
3. Plant height (m): 2.3
4. Plant girth (mm): 14.6
5. Biomass yield (t ha^{-1}): 38.1
6. Juice yield (t ha^{-1}): 15.8
7. Juice extraction (%): 41.5
8. Brix (%): 17
9. Sugar yield (t ha^{-1}): 2.0
10. Grain yield (t ha^{-1}): 3.2
11. Male fertility restoration (%): 85
12. Sucrose (%): 6.9
13. Glucose (%): 1.1
14. Fructose (%): 1.5
15. pH of juice: 5.2
16. Electrical conductivity of juice (mS m^{-1}—milli siemens per metre): 9.6
17. Tolerance to: Aphids, Shoot fly, Anthracnose
18. Adaptation: Post-rainy season

Traits recorded as per guidelines for sorghum as approved by PPVFRA[a]

Characteristics	Characteristic value of candidate hybrid									Remarks measured value etc.
	1	2	3	4	5	6	7	8	9	
Seedling: anthocyanin colouration of coleoptile	√									Purple
Leaf sheath: anthocyanin colouration		√								Purple
Leaf: midrib colour (5th fully developed leaf)	√									Dull green
Plant: time of panicle emergence (50 % of the plants with complete panicle emergence)								√		78
Plant: natural height of foliage up to base of flag leaf			√							2.0 m
Flag leaf: extension of discoloration of midrib	√									Absent
Flag leaf: intensity of green coloration of midrib compared to the blade		√								Same color
Flag leaf: yellow coloration of midrib	√									Absent
Glume hair color	√									Absent
Lemma: arista formation (awns)	√									Absent
Stigma: anthocyanin coloration	√									Absent
Stigma: yellow coloration					√					Medium
Stigma length (mm)			√							Short
Flower with pedicel: length of flower			√							Short
Anther: length			√							Short
Anther: colour of dry anther				√						Orange
Glume: color		√								Straw
Plant: total height					√					2.3 m
Stem: diameter (at lower one-third height of plant) (mm)			√							14.6
Leaf: length of blade of the third leaf from top including flag leaf (cm)							√			65.78
Leaf: width of blade of the third leaf from top including flag leaf (cm)							√			7.61
Panicle: length without peduncle			√							17.78
Panicle: length of branches (middle third of panicle)				√						6.67
Panicle: density at maturity (ear head compactness)							√			Semi compact
Panicle: shape			√							Symmetric
Neck of panicle: visible length above sheath (cm)					√					15.33
Glume coverage (%)	√									25
Shattering			√							Low
Threshability				√						Partly threshable
Grain form	√									Single
Caryopsis: colour after threshing			√							Yellow
Grain: weight of 1000 grains (g)		√								26.0
Grain: shape in dorsal view		√								Circular
Grain: shape in profile view		√								Circular
Grain: size of mark of germ				√						Medium
Grain: texture of endosperm (in longitudinal section)				√						50 % corneous
Grain: colour of vitreous albumen			√							Yellow
Grain: lustre	√									Non-lustrous
Seedling vigor score	√									Most vigorous
Leaf glossy score	√									Glossy
Plant aspect score		√								Average

[a] *PPVFRA* Protection of Plant Varieties and Farmers Rights Act

16 ICSSH 76 Salient Features

1. Pedigree: ICSA 502 × ICSV 25284
2. Days to 50 % flowering: 80 days
3. Plant height (m): 1.9
4. Plant girth (mm): 13.6
5. Biomass yield (t ha^{-1}): 34.4
6. Juice yield (t ha^{-1}): 14.2
7. Juice extraction (%): 41.4
8. Brix (%): 13
9. Sugar yield (t ha^{-1}): 1.4
10. Grain yield (t ha^{-1}): 4.2
11. Male fertility restoration (%): 97
12. Sucrose (%): 3.3
13. Glucose (%): 1.1
14. Fructose (%): 1.3
15. pH of juice: 5.4
16. Electrical conductivity of juice (mS m^{-1}—milli siemens per metre): 12.9
17. Tolerance to: Anthracnose
18. Adaptation: Post-rainy season

Traits recorded as per guidelines for sorghum as approved by PPVFRA[a]

Characteristics	Characteristic value of candidate hybrid									Remarks measured value etc.
	1	2	3	4	5	6	7	8	9	
Seedling: anthocyanin colouration of coleoptile	√									Green
Leaf sheath: anthocyanin colouration	√									Tan
Leaf: midrib colour (5th fully developed leaf)		√								Dull green
Plant: time of panicle emergence (50 % of the plants with complete panicle emergence)										80
Plant: natural height of foliage up to base of flag leaf			√							1.7 m
Flag leaf: extension of discoloration of midrib	√									Absent
Flag leaf: intensity of green coloration of midrib compared to the blade		√								Same color
Flag leaf: yellow coloration of midrib	√									Absent
Glume hair color	√									Absent
Lemma: arista formation (awns)	√									Absent
Stigma: anthocyanin coloration	√									Absent
Stigma: yellow coloration					√					Medium
Stigma length (mm)			√							Short
Flower with pedicel: length of flower					√					Medium
Anther: length					√					Medium
Anther: colour of dry anther				√						Orange
Glume: color		√								Straw
Plant: total height					√					1.9 m
Stem: diameter (at lower one-third height of plant) (mm)					√					13.6
Leaf: length of blade of the third leaf from top including flag leaf (cm)							√			78.11
Leaf: width of blade of the third leaf from top including flag leaf (cm)					√					8
Panicle: length without peduncle					√					25
Panicle: length of branches (middle third of panicle)					√					8.22
Panicle: density at maturity (ear head compactness)					√					Semi loose
Panicle: shape			√							Symmetric
Neck of panicle: visible length above sheath (cm)								√		23.78
Glume coverage (%)				√						33.33
Shattering			√							Low
Threshability					√					Partly threshable
Grain form	√									Single
Caryopsis: colour after threshing			√							Yellow
Grain: weight of 1000 grains (g)					√					31.0
Grain: shape in dorsal view			√							Circular
Grain: shape in profile view			√							Circular
Grain: size of mark of germ					√					Medium
Grain: texture of endosperm (in longitudinal section)					√					50 % corneous
Grain: colour of vitreous albumen			√							Yellow
Grain: lustre					√					Medium
Seedling vigor score	√									Most vigorous
Leaf glossy score	√									Glossy
Plant aspect score			√							Average

[a] PPVFRA Protection of Plant Varieties and Farmers Rights Act

17 CSH 22 SS Salient Features

1. Pedigree: ICSA 38 × SSV 84
2. Days to 50 % flowering: 73 days
3. Plant height (m): 2.1
4. Plant girth (mm): 13.8
5. Biomass yield (t ha^{-1}): 36.7
6. Juice yield (t ha^{-1}): 13.1
7. Juice extraction (%): 35.6
8. Brix (%): 13
9. Sugar yield (t ha^{-1}): 1.3
10. Grain yield (t ha^{-1}): 3.7
11. Male fertility restoration (%): 100
12. Sucrose (%): 6.4
13. Glucose (%): 1.8
14. Fructose (%): 1.0
15. pH of juice: 5.3
16. Electrical conductivity of juice (mS m^{-1}—milli siemens per metre): 18.0
17. Tolerance to: Aphids, Shoot fly
18. Adaptation: Post-rainy season

Traits recorded as per guidelines for sorghum as approved by PPVFRA[a]

Characteristics	Characteristic value of candidate hybrid									Remarks measured value etc.
	1	2	3	4	5	6	7	8	9	
Seedling: anthocyanin colouration of coleoptile	√									Green
Leaf sheath: anthocyanin colouration	√									Tan
Leaf: midrib colour (5th fully developed leaf)		√								Dull green
Plant: time of panicle emergence (50 % of the plants with complete panicle emergence)					√					73
Plant: natural height of foliage up to base of flag leaf			√							1.8 m
Flag leaf: extension of discoloration of midrib					√					Medium
Flag leaf: intensity of green coloration of midrib compared to the blade		√								Same color
Flag leaf: yellow coloration of midrib	√									Absent
Glume hair color	√									Absent
Lemma: arista formation (awns)	√									Absent
Stigma: anthocyanin coloration	√									Absent
Stigma: yellow coloration					√					Medium
Stigma length (mm)					√					Medium
Flower with pedicel: length of flower					√					Medium
Anther: length					√					Medium
Anther: colour of dry anther				√						Orange
Glume: color					√					Red
Plant: total height					√					2.1 m
Stem: diameter (at lower one-third height of plant) (mm)					√					13.8
Leaf: length of blade of the third leaf from top including flag leaf (cm)						√				63.87
Leaf: width of blade of the third leaf from top including flag leaf (cm)							√			7.56
Panicle: length without peduncle					√					23.44
Panicle: length of branches (middle third of panicle)					√					8.17
Panicle: density at maturity (ear head compactness)			√							Loose
Panicle: shape			√							Symmetric
Neck of panicle: visible length above sheath (cm)								√		26.22
Glume coverage (%)	√									25
Shattering			√							Low
Threshability					√					Partly threshable
Grain form	√									Single
Caryopsis: colour after threshing				√						Yellow
Grain: weight of 1000 grains (g)			√							25.0
Grain: shape in dorsal view			√							Circular
Grain: shape in profile view			√							Circular
Grain: size of mark of germ					√					Medium
Grain: texture of endosperm (in longitudinal section)							√			25 % corneous
Grain: colour of vitreous albumen			√							Yellow
Grain: lustre	√									Non-lustrous
Seedling vigor score	√									Most vigorous
Leaf glossy score	√									Glossy
Plant aspect score		√								Good

[a] PPVFRA Protection of Plant Varieties and Farmers Rights Act

Commercialization: Status and Way Forward

P. Srinivasa Rao, C. Ganesh Kumar and Belum V. S. Reddy

Abstract Ethanol is a biofuel that is used as a fuel additive and a replacement for nearly 3 % of the world's fossil fuel-based gasoline consumption. Currently, most of the bioethanol is produced from sugarcane in Brazil and corn in the United States, while biodiesel is made from rapeseed in Europe. The rationale for the success of the Brazilian Proalcool program, its present status and its perspectives has been presented. The Proalcool program's mandate was a vast increase in ethanol production with a sound government-backed subsidies and incentives initially to reach the goal; however, it was the private investors and companies that were solely responsible to achieve the end result. The Proalcool program indeed provides several essential lessons to many countries around the world about the potential competitiveness of biofuels *vis-à-vis* traditional fuels. Considering the importance of alternate biofuels, sweet sorghum has been identified as a promising energy crop to meet the energy security and reduce the dependence on fossil fuels in many countries around the globe. The Indian National Biofuel Policy (2009) recognizes sweet sorghum as a major biofuel feedstock and well adapted to India. However, its value chain could not get popular as anticipated due to low price level (Rs. 27) fixed by Government of India. Hence, it is necessary to review the ethanol price in India so as to give fillip to the beleaguered biofuel industry, which will likely to play a stabilization role in a oil import dependent economy like ours. Similarly, a number of case studies are presented on the research efforts made in various countries around the world like India, USA, Brazil and China on the use of sweet sorghum as

P. Srinivasa Rao (✉) · B. V. S. Reddy
International Crops Research Institute for the Semi-Arid Tropics (ICRISAT),
Patancheru 502324, India
e-mail: psrao72@gmail.com

C. Ganesh Kumar
Chemical Biology Laboratory, CSIR-Indian Institute of Chemical Technology (CSIR-IICT),
Uppal Road, Hyderabad 500607, India

P. S. Rao and C. G. Kumar (eds.), *Characterization of Improved Sweet Sorghum Cultivars*, SpringerBriefs in Agriculture,
DOI: 10.1007/978-81-322-0783-2_5, © The Author(s) 2013

a potential bioenergy feedstock. The current and commercialization status of the various biofuel technologies and approaches are discussed. The biofuel blending targets and mandates of different countries are also presented.

Keywords Commercialization · Biofuel blending targets · Brazil · Sweet sorghum · Proalcool · Sugarcane · Manioc · Distilleries · Renewable · Bioethanol · Genetic improvement · Sorganol · Feedstocks

1 Overview

Bioenergy has become a priority area for research and development worldwide and nations are investing heavily to increase their energy security and reduce their fossil-fuel carbon emissions and pollution. More than 50 countries, including several developing countries, have adopted blending targets or mandates and several more have announced biofuel quotas for future years (Table 1).

As a result now biofuels provide around 3 % of total road transport fuel globally (on an energy basis) and considerably higher shares are achieved in certain countries and its share is expected to reach 27 % of energy basket by 2050. Brazil, currently, met about 21 % of its road transport fuel demand in 2008 with biofuels. Similarly, in the United States, the share was 4 % of road transport fuel and in the European Union (EU) around 3 % in 2008. For this, a wide variety of conventional and advanced biofuel conversion technologies exists today. The current status of the various technologies and approaches to biofuel production is given in Fig. 1. Conventional biofuel processes, though already commercially available, continue to improve in efficiency and economics. Advanced conversion/ processing technologies are moving to the demonstration stage or are already there (e.g., biomethane/syngas production).

2 The Success Story: Brazil

In Brazil, the sugarcane production industry has historically been concentrated in two main areas in the country, the Northeast, in the states of Algoas and Pernambuco and the Center South, in the state of Sao Paulo, where a large number of sugarcane plantations were owned and operated by small, independent farmers (Bolling and Suarez 2001). Later, the Brazilian government launched its biofuel initiative on 14 November 1975 by the presidential decree number 76.593, the PROALCOOL (*Programa Nacional do Alcool*) in response to the worldwide oil crisis in 1973 and to look possible domestic sources for alternative fuel production in order to insulate itself from the chaotic market (Cordonnier 2008). The program was aimed at bolstering Brazil's national sugar economy by safeguarding the

Table 1 Overview of biofuel blending targets and mandates

Country/region	Current mandate/ target	Future mandate/target	Current status (mandate [M]/ target [T])
Argentina	E5, B7	n.a.	M
Australia: New South Wales (NSW), Queensland (QL)	NSW: E4, B2	NSW: E6 (2011), B5 (2012); QL: ES (on hold till autumn 2011)	M
Bolivia	E10, B2.5	B20 (2015)	T
Brazil	E20-25, B5	n.a.	M
Canada	E5 (up to E8.5 in 4 provinces), B2-B3 (in 3 provinces)	B2 (nationwide) (2012)	M
Chile	E5, B5	n.a.	T
China (9 provinces)	E10 (9 provinces)	n.a.	M
Colombia	E10, B10	B20 (2012)	M
Costa Rica	E7, B20	n.a.	M
Dominican Republic	n.a.	E15, B2 (2015)	n.a.
European Union	5.75 % biofuels[a]	10 % renewable energy in transport[b]	T
India	E5	E20, B20 (2017)	M
Indonesia	E3, B2.5	E5, B5 (2015); E15, B20 (2025)	M
Jamaica	E10	Renewable enegy in transport: 11 % (2012); 12.5 % (2015); 20 % (2030)	M
Japan	500 Ml/y (oil equivalent)	800 Ml/y (2018)	T
Kenya	E10 (in Kisumu)	n.a.	M
Korea	B2	B2.5 (2011); B3 (2012)	M
Malaysia	B5	n.a.	M
Mexico	E2 (in Guadalajara)	E2 (in Monterrey and Mexico City; 2012)	M
Mozambique	n.a.	E10, B5 (2015)	n.a.
Norway	3.5 % biofuels	5 % proposed for 2011; possible alignment with EU mandate	M
Nigeria	E10	n.a.	T
Paraguay	E24, B1	n.a.	M
Peru	E7.8, B2	B5 (2011)	M
Philippines	E5, B2	B5 (2011), E10 (Feb. 2012)	M
South Africa	n.a.	2 % (2013)	n.a.
Taiwan	B2, E3	n.a.	M
Thailand	B3	3 Ml/d ethanol, B5 (2011); 9 Ml/d ethanol (2017)	M
Uruguay	B2	E5 (2015), B5 (2012)	M

(continued)

Table 1 (continued)

Country/region	Current mandate/ target	Future mandate/target	Current status (mandate [M]/ target [T])
United States	48 billion liters of which 0.02 bln. cellulosic-ethanol	136 billion liters, of which 60 bln. cellulosic-ethanol (2022)	M
Venezuela	E10	n.a.	T
Vietnam	n.a.	50 Ml biodiesel, 500 Ml ethanol (2020)	n.a.
Zambia	n.a.	E5, B10 (2011)	n.a.

B: biodiesel (B2: 2 % biodiesel blend); E: ethanol (E2: 2 % ethanol blend); Ml/d: million liters per day. n.a.: not available
[a] Currently, each member state has set up different targets and mandate
[b] Lignocellulosic biofuels, as well as biofuels made from wastes and residues, count twice and renewable electricity 2.5-times towards the target
Source International Energy Agency (2010) analysis based on various governmental sources. For more information see also: http://renewables.iea.org

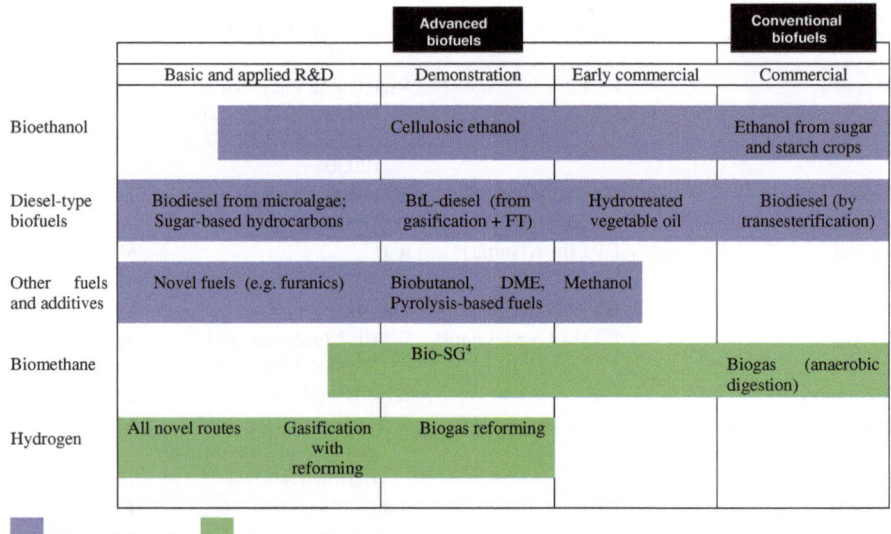

Liquid biofuel   Gaseous biofuel
1. Biomass-to-liquids; 2. Fischer-Tropsch; 3. Dimethyl ether; 4. Bio-synthetic gas

Fig. 1 Commercialization status of main biofuel technologies (*Source* Modified from Bauen et al. 2009)

privately owned sugar industry and to substitute the petroleum oil imports with that of locally produced ethanol by converting surplus sugar into anhydrous ethanol. The program implemented and regulated the use of hydrated ethanol as fuel by blending it with petroleum gasoline. Article 2 of the Proalcool decree

Fig. 2 Moema sugar mill located in Orindiuva, Sao Paulo, Brazil, producing transport-grade biofuel (*Source* Moraes 2011)

allowed the use of either sugarcane or manioc root for the production of ethanol and the specific mandate was to produce 3.5 billion liters of ethanol from sugarcane by 1980. Proalcool offered subsidies to both manioc-based plants and sugar cane-based plants (Fig. 2); the de facto standard in the alcohol industry was to establish plants producing 120,000 to 240,000 l/day, a scale that far exceeded the typically smaller manioc plants. The price paid to producers in 1980 was US$ 700 for 1,000 l; over the years with the gains in the technology and economics of scale has driven the production costs down, reaching as low as US$ 200 per 1,000 l in 2004. In the past thirty five years, the ethanol industry has expanded enormously due largely to strong governmental incentives, subsidies, mandates and pro-ethanol legislation. Demetricus (1990) disputed that low agricultural productivity and production led most investors to choose sugarcane as the preferred raw material for ethanol production. The Proalcool program's mandate was a vast increase in ethanol production with a sound government-backed subsidies and incentives to reach that goal; however, it was the private investors and companies that were solely responsible to achieve the end result. In the course of time, the Brazilian ethanol market experienced a phase of rapid expansion stimulated by the development of flex fuel vehicles (FFVs). The ethanol content in these blends started initially at 5 % and the current Brazil's government mandates has increased to up to 25 % blending (currently even up to E85-E100 in FFVs) in gasoline since the last thirty five years of Proalcool. The history of ethanol production in Brazil provides an interesting insight into how an authoritarian regime, dedicated to promote an alternative renewable fuel, managed to achieve that goal. The Proalcool program indeed provides several essential lessons to many countries around the world about the potential competitiveness of biofuels *vis-à-vis* traditional fuels. Brazil with a vibrant biofuel industry has about 437 ethanol producing plants and a typical plant crushes about 2 million tons of sugarcane per year (Goldemberg

2008) and annually turns half of its sugarcane harvest into 20 billion liters of ethanol to power 12.5 million vehicles (45 % of Brazilian vehicles). Majority of the large plants are located in the state of Sao Paulo where almost two-thirds of the ethanol is being produced. Today Brazil is the highest producer of ethanol in world after USA and the government intervention is basically limited to determining the proportion of the anhydrous ethanol blend, setting the tax rate on sugar exports, etc.

3 The Sweet Sorghum Story

Research experiences gained on the cultivation of sweet sorghum in India, USA, Brazil and China have shown that the crop has high potential as a bio-energy feedstock, with several opportunities for immediate use as a complementary feedstock in dry and semi-arid land pockets and as a seasonal low-cost feedstock (mold-affected grain). Regions with a warm climate, large tracts of land and a system similar to sugarcane processing should work well for sweet sorghum. Further, the crop can be grown in regions of the world where sugarcane cannot be cultivated. The required government policy support is necessary for utilization of this novel feedstock for commercial bioethanol production. Across the globe a handful of distilleries started using sweet sorghum for ethanol production on commercial scale since 2007. M/s. Rusni Distilleries Pvt. Ltd. is the first sweet sorghum distillery established in 2007 at Sangareddy, Medak district of Andhra Pradesh, India and is amenable to multiple feedstocks. It has a production capacity of 40 kl/day (KLPD) and produces fuel ethanol (99.6 % alcohol), Extra Neutral Alcohol (ENA, 96 %) and pharma alcohol (99.8 %) from different agro-based raw materials such as sweet sorghum stalks (juice), molded grains, broken rice, cassava and rotten fruits. Another, 30 KLPD Tata Chemicals Limited distillery located in Nanded, Maharashtra, started operations in 2009 solely based on sweet sorghum and produced 90 KL of transport grade ethanol in 2010. However, both the distilleries stopped operations primarily due to low market price of ethanol (Rs. 27 or 50 cents per litre).

The Chinese bioethanol production cost from corn is equivalent to 1.022 US$/liter, while the U.S. bioethanol production cost from corn was 0.492 US$/liter (Licht 2008). The feedstock cost of sweet sorghum was 2,000 Yuan/ton, while the bioethanol production cost from sweet sorghum was 4,400 Yuan/ton (Song et al. 2008). In view of the high feedstock prices, the Chinese government is providing subsidies to cover the operating costs. The average subsidy for fuel bioethanol production set by the Chinese government was 1,836 Yuan/ton in 2005, 1,625 Yuan/ton in 2006, 1,374 Yuan/ton in 2007, and 1,754 Yuan/ton in 2008. Further, it was estimated that the removal of Value Added Tax and Consumption Tax which totaled to about 190 million Yuan (US$28 million), and the direct financial subsidy totaled to about 2 billion Yuan (US$294 million) for grain-based bioethanol plants from 2002 to 2008 (Lang et al. 2009). In January 2006, the Chinese government enacted the "Renewable Energy Law" to promote renewable energy utilization and production. Saline alkaline soils were preferred in a number of Chinese provinces for cultivating sweet sorghum; however, the production status was much lower as compared to corn production. In 2010–2011, the Chinese sweet sorghum production was 1.5 million tons and corn production was 28.6 million tons (USDA-FAS 2011). In addition, the Chinese sweet sorghum-bioethanol production was technically immature and bioethanol content was so low (20 %) that it could not be used as fuel (Wang 2011). At present, the status of biofuel production from sweet sorghum in China is still in the pilot scale project stage. Different biomass producing companies based in China such as Liaoning Guofu Bioenergy Development Company Limited, Binzhou Guanghua Biology Energy Company Ltd, Jiangxi Qishengyuan Agri-Biology Science and Technology Company Ltd, Xinjiang Santai Distillery, Jilin Fuel Alcohol Company Limited, Heilongjiang Huachuan Siyi Bio-fuel Ethanol Company Ltd, ZTE Agribusiness Company Limited and Fuxin Green BioEnergy Corporation—have conducted large-scale sweet sorghum trials. However, a few problems were identified in the processing of sweet sorghum stalks. In 2010, ZTE Agribusiness Company Limited, Wuyuan County, Inner Mongolia and Fuxin Green BioEnergy Corporation, Heishan County, Shenyang province used sweet sorghum as feedstock to produce ethanol. The Chinese government was encouraging sweet sorghum processing industries by offering a subsidy of ¥180 mu^{-1} to farmers or companies cultivating sweet sorghum and ¥1300 t^{-1} for ethanol produced to the industry. Given these sops, the area under sweet sorghum is likely to increase substantially in the near future.

In the Philippines, San Carlos Bio-Energy Incorporated set up the first commercial bioethanol distillery for fuel production at Visyan Islands of Negro. This firm used the sweet sorghum variety SPV 422 developed at ICRISAT-Patancheru and 14 K fuel grade ethanol was produced in 2012 from sweet sorghum syrup (247 l of fuel grade ethanol from a ton of syrup). In addition, it used sugarcane to extend the operation of the distillery during the offseason. Further, several upcoming sugarcane distilleries in the provinces of Bukidnon (Mindanao), Tarlac and Pampanga (Luzon), which have large tracts of idle land suitable for sweet sorghum cultivation, are exploring the possibility of using sweet sorghum as a complementary feedstock. In Bicol region of Philippines, a development program on the commercialization of sweet sorghum products and by-products was implemented through public–private sector partnership. At Batac, Bapamin enterprises lead by Antonio Arcangel is marketing vinegar (1,000 l/month) and other food products from sweet sorghum commercially since 2008 (Reddy et al. 2011).

In USA, EnviroFuels, LLC, Riverview, Florida is currently in the process of developing a 30 million gallon per year sugar-based Advanced Biofuel ethanol plant in Highlands County, Florida using sweet sorghum as the primary feedstock. Sugarcane will be used to supplement the feedstock base in the winter months when sweet sorghum is not available. Another firm, Southeast Renewable Fuels LLC is building a 20 million gallon per year sweet sorghum-to-ethanol advanced bio-refinery in Hendry County, near Clewiston, Florida. BioDimensions Industrial Sugar Platform Development, located in the state of Tennessee, USA, planted and harvested about 75 ha of sweet sorghum. A portion of the sugars was fermented to ethanol, while the bagasse was used for making both fuel pellets and animal feed. Energy sugar beets are the complementary off-season feedstock being used when sweet sorghum is not available. The group expects to distill about 5,000 gallons of hydrous ethanol, much of which will be used in an industrial ethanol engine.

The Ceres, Inc. established a subsidiary in Brazil is focusing on developing sweet sorghum as a feedstock for the ethanol industry. The company's goal is to be the first supplier of new hybrids with high levels of sugar. The company is currently working with multiple ethanol mills, technology providers and equipment companies to facilitate the introduction of sweet sorghum hybrids into existing ethanol mills. The EMBRAPA is also closely working with several sugar mills and produced over 5,000 l of transport grade ethanol on pilot scale basis in Sao Paulo region and has already indented 200 tonnes of seed for its sweet sorghum variety, BRS 506. It is anticipated that in the coming years, sweet sorghum will become the second most important bioethanol feedstock in Brazil after sugarcane, optimally exploiting the season between two sugarcane crops.

Net energy ratio and greenhouse gas balance primarily decides the benefits of an energy value chain. However, it was reported that sweet sorghum has a high net energy balance of 3.63 as compared to grain sorghum (1.50) and corn (1.53) balance (Wortmann et al. 2008). Another report estimated an energy balance of 8 and carbon emission reduction by 86 % (CII-DBT Report 2010). First and second generation bioethanol from sweet sorghum can contribute significantly to the conservation of fossil resources and to the mitigation of greenhouse gases. If the crop is used for the production of ethanol (from grains and sugar) and green electricity (from surplus bagasse), 3,500 l of crude oil equivalents can be saved per hectare cultivation area. If both food from grains and ethanol from the juice are produced, 2,300 l of crude oil equivalents can be saved per hectare cultivated area. Regarding greenhouse gases, between 1.4 and 22 kg CO_2 equivalents can be saved depending on yield, production methods and the land cover prior to sweet sorghum cultivation (Köppen et al. 2009). For both categories, the exact values vary greatly with specific scenarios and local conditions. In general, the following parameters that determine the results are the type and efficiency of conversion technology, the use of byproducts (e.g., bagasse), the crop yield per cultivation area, land-use changes, as well as the type of fossil energy carriers that are replaced. Even if the seeds were used as food, bioethanol from the stem's sugar juice still shows clear advantages over fossil fuels. If both sugar and seeds were used as food, the respective conversion related energy and greenhouse gas expenditures could be compensated by producing second generation ethanol from the bagasse. Even though the ethanol yield per unit weight of feedstock is lower for sweet sorghum as compared to sugarcane, the much lower production costs and water requirement for this crop more than compensates for the difference, and hence, it still returns a competitive cost advantage for the production of ethanol in India (Farrell et al. 2006).

There are many factors affecting the sweet sorghum value chain. The following major challenges identified are:

- G × E interactions were significant for sweet sorghum related traits; the genotypes that performed well in the rainy season were not necessarily the top-performers in the post-rainy season and *vice versa.* Preliminary results indicated that non-allelic interactions were more predominant for stalk sugar and allied traits.
- As global climate is gradually changing to higher temperatures and sweet sorghum is bound to grow in new areas, thermo- and photo-insensitive non-lodging cultivars that are resistant to multiple pests and diseases needs to be developed.
- Sugars accumulation and sustenance is a complex process and is governed by many alleles. Cool season or temperate sweet sorghums need to be evolved.
- Breeding of short, mid-late and late maturing genotypes is necessary in order to have a broad harvest window in sweet sorghum, and thus providing raw material to the distillery over a long period. Proper planning of sowing of a mixture of these cultivars in the catchment area of distillery would help in achieving more commercial stalk sugar/ethanol.
- When cultivars with different maturity groups are grown in an area, pests like shootfly and midge are prone to infest the late maturing cultivars. Therefore, breed for those insect-tolerant cultivars.
- Sorghum crop is traditionally challenged by marginal lands with poor fertility status and poor moisture holding capacity and sweet sorghum too encounters similar problems. Sporadic water inundation due to excessive rains/floods also becomes an unforeseen constraint.
- The self fermentation of juice inside the stalk prior to juice extraction is a major concern, mainly when juice extraction is delayed after harvesting due to long distances prevailing between factory and the field. Preliminary results indicated that there will be reduction in sugar yield by 16.8 %, if the juice extraction is delayed by 48 h (Srinivasarao et al. 2012). Research should therefore address the problem of post-harvest losses in terms of juice quality and quantity.

4 Way Forward

Sweet sorghum has a low water demand and is especially advantageous in areas with water shortage (Srinivasarao et al. 2011). Its lower nitrogen fertilizer demand possibly due to traits such as biological nitrification inhibition (BNI), reduces the risk of nutrient leaching and thus soil and water pollution, as well as making it well suited for small-scale farming. Its relatively short vegetation cycle allows sweet sorghum to be grown in double cropping systems which involves the harvesting of the crop twice or more number of times from a single planting during the growing season (Duncan and Gardner 1984) based on water availability, which in turn can lead to greater agro-biodiversity and a reduced demand for fertilizers and pesticides.

A limiting factor for its widespread cultivation is the limited availability of varieties/hybrids adapted to different agro-climatic conditions (e.g., lack of post-rainy season adapted lines in India) resisting both biotic and abiotic stresses, including colder climate. Consequently, research should address the optimization of sweet sorghum as an energy crop through breeding for enhanced productivity under limited available resources. Genetic improvement should focus on stalk sugar, biomass quantity and quality and in general, agronomic traits (such as water and nutrient use efficiency) and in particular, adaptation of sweet sorghum to colder, arid saline, and alkaline conditions. Further, an improvement in Brix%, juice volume and stalk yield ($\geq$45 t ha^{-1} with hybrids) should be targeted in sweet sorghum to help improve the benefits to the industry and farmers without any detrimental effect on grain yield. The juice volume should not be compromised, while increasing the Brix%. There is also a need to develop and evaluate cultivars producing high stalk yield per unit time, input, energy and land area in different agro-climatic regions of the country. Other research areas on quality and processing which needs immediate attention include high ethanol yield, fermentation efficiency, diffusion, diversified products from bagasse (power, pulp, bio-manure, cattle feed, etc.).

As the demand for biofuels rapidly expands, its associated production systems and supply chains are consolidating. Forward-thinking management systems could significantly enhance ecological sustainability and livelihood development, particularly for poor farmers in the developing world. International trade will be crucial to enlarge the share of bioethanol in future transport energy demand. In the longer terms, developing countries can profit from the experiences with sustainable conventional biofuel production (e.g., Brazil and USA) and later adopt advanced biofuel technologies once they are commercially proven. If the countries with sound policy framework targeting the entire sweet sorghum innovation chain to ensure that the development and use of biofuels, in general and sorganol in particular, would reap rich dividends in climate change mitigation and adaptation, energy security and all round sustainable economic development, without compromising food or feed security. Full exploration of the available genetic resources through plant breeding with the aid of molecular tools could dramatically increase biomass yield of sorghum and thus meet the demand of feedstocks for biofuel production without a significant impact on our food supply and natural environment.

Acknowledgments We would like to thank the donor agencies—International Fund for Agriculture Development (IFAD), National Agriculture Innovation Project (NAIP-ICAR), and European Commission-SWEETFUEL (EC) for funding projects on sweet sorghum improvement and value chain development to further the cause of SAT farmers. The efforts of Mr. K Devendram, Mr. B. Ramaiah, Mr. Ch. Madhu (ICRISAT); Dr. R. Nageswar Rao, Dr. M. Vairamani and Mrs. Sara Khalid (IICT) in field and lab data collection and analysis are also acknowledged.

References

Bauen A, Berndes G, Junginger M, Londo M, Vuille F (2009) Bioenergy—a sustainable and reliable energy source. A review of status and prospects. IEA Bioenergy: ExCo:2009:06

Bolling C, Suarez NR (2001) The Brazilian sugar industry: recent developments, sugar and sweetener situation & outlook/SSS-232 (Sept. 2001) (Economic Research Service/USDA). http.//www.ers.usda.gov/BriefingBrazil/braziiansugar.pdf

CII-DBT Report (2010) Estimation of energy and carbon balance of biofuels in India, Feb 2010
Cordonnier VM (2008) Ethanol's roots: how Brazilian legislation created the international
    ethanol boom. William Mary Environ Law Policy Rev 33:287–317
Demetricus FJ (1990) Brazil's national alcohol program: technology and development in an
    authoritarian regime. Praeger Publishers, New York, 181 pp
Duncan RR, Gardner WA (1984) The influence of ratoon cropping on sweet sorghum yield, sugar
    production, and insect damage. Can J Plant Sci 64:261–273
Farrell AE, Plevin RJ, Turner BT, Jones AD, O'Hare M, Kammen DM (2006) Ethanol can
    contribute to energy and environmental goals. Science 311(5760):506–508
Goldemberg J (2008) The Brazilian biofuels industry. Biotechnol Biofuels 1:6. doi:10.1186/
    1754-6834-1-6
International Energy Agency (2010) Sustainable production of second generation bio-fuels:
    potential and perspectives in major economies and developing countries
Köppen S, Reinhardt G, Gärtner S (2009) Assessment of energy and greenhouse gas inventories
    of sweet sorghum for first and second generation bioethanol. FAO Environmental and Natural
    Resources Service Series, No. 30, FAO, Rome, 83 pp
Lang X, Zheng F, Cui H (2009) Evolution of fuel ethanol policy in China. For Econ, pp 29–33
Licht FO (2008) Ethanol production costs a worldwide survey, F.O. Licht, Ratzeburg, Germany.
    ISSN 14785765
Moraes M (2011) Lessons from Brazil. Nature 474:S25–S25
Reddy BVS, Layaoen H, Dar WD, Srinivasa Rao P, Eusebio JE (Eds.) (2011) Sweet sorghum in
    the Philippines: status and future. International Crops Research Institute for the Semi-Arid
    Tropics, Patancheru 502 324, Andhra Pradesh, India. 116 pp
Song A, Pei G, Wang F, Wan D, Feng C (2008) Survey for fuel biofuel feedstock multiple
    production. Acad Report Agric Process 24:302–307
Srinivasarao P, Sanjana Reddy P, Rathore A, Reddy BVS, Sanjeev P (2011) Application of GGE
    biplot and AMMI model to evaluate sweet sorghum (Sorghum bicolor) hybrids for
    genotype × environment interaction and seasonal adaptation. Indian J Agric Sci 81:438–444
Srinivasarao P, Kumar CG, Malapaka J, Kamal A, Reddy BVS (2012) Feasibility of sustaining
    sugars in sweet sorghum stalks during post-harvest stage by exploring cultivars and
    chemicals: a desk study. Sugar Tech 14:21–25
USDA-FAS (U.S. Department of Agriculture, Foreign Agricultural Service) (2011) PS&D,
    USDA-FAS, 01.03.2011. http://www.fas.usda.gov/psdonline/psdQuery.aspx
Wang QA (2011) Time for commercializing non-food biofuels in China. Renew Sustain Energy
    Rev 15:621–629
Wortmann C, Ferguson R, Lyon D (2008) Sweet sorghum as a biofuel crop in Nebraska. Paper
    presented at the 2008 Joint Annual Meeting, Celebrating the International Year of Planet
    Earth, 5–9 Oct 2008, Houston, Texas. http://crops.confex.com/crops/2008am/techprogram/
    P44581.htm

About the Editors

Dr. P. Srinivasa Rao is a Senior Scientist working in the area of sorghum breeding at the International Crops Research Institute for the Semi-Arid Tropics (ICRISAT), Patancheru, Hyderabad, India. His current areas of research interest are on biofuels, resistance breeding, brown midrib mutants, energy and sweet sorghums, sucrose transporters as well as wide hybridization. He obtained his Ph.D. (Genetics) 1998 from Indian Agricultural Research Institute (IARI), New Delhi. He is a recipient of DBT- postdoctoral fellowship (2005-2008) and worked on interactions of bacterial leaf blight pathogen with R-gene pyramids of Rice with Dr. Ramesh V Sonti at Centre for Cellular and Molecular Biology (CCMB), Hyderabad, India. He served as Scientist (Plant Breeding) for seven years in Acharya NG Ranga Agricultural University (ANGRAU), Hyderabad, India. He is a recipient of the Distinguished Researcher Award (2011) from Pampanga Agricultural College, Philippines. He is serving as an editorial board member of Sugar Tech. and Rangeland Science. He has 28 research publications in various national and international journals, 6 reviews, 15 book chapters and 3 books to his credit.

Dr. C. Ganesh Kumar is a Senior Scientist working in the area of chemical biology with the CSIR-Indian Institute of Chemical Technology, Hyderabad, India. His current areas of research interest are on biofuels, bioactives, biosurfactants, enzymes and biotransformations. He obtained his Ph.D. (Microbiology) 1997 from National Dairy Research Institute (ICAR), Karnal, Haryana under the guidance of Prof. M.P. Tiwari. During his Ph.D. tenure, he also received a DAAD short-term fellowship and worked with Prof. K.-D. Jany, Director and Head, Centre for Molecular Biology, Federal Research Centre for Nutrition, Karlsruhe, Germany. He worked as a post-doctoral fellow at various reputed organizations like Department of Biochemistry, Bose Institute, Kolkata, India; Department of Biochemistry, College of Medicine, Inha University, Inchon, South Korea with Prof. Chung-Soon Chang and at the Bioproduction Research Institute, National Institute of Advanced Industrial Science and Technology (AIST), Sapporo, Japan, with Prof. Isao Yumoto as a JSPS fellow on different

P. S. Rao and C. G. Kumar (eds.), *Characterization of Improved Sweet*
Sorghum Cultivars, SpringerBriefs in Agriculture,
DOI: 10.1007/978-81-322-0783-2, © The Author(s) 2013

extremophiles and their metabolites. He is a recipient of the AMI Young Scientist Award (1999) from the Association of Microbiologists of India, Fellow (2009) of the Association of Biotechnology and Pharmacy, Eminent Bioengineer Award (2010) from the Society for Applied Biotechnology and IICT Gaurav Samman Award (2010). He has 45 research publications in various national and international journals, 10 reviews, 2 book chapters and 3 patents to his credit.